Virtual Machine based Mechanisms and To[ols for] Attack Prevention, Analysis, and Recovery

By

DANIELA ALVIM SEABRA DE OLIVEIRA
B.S. (Federal University of Minas Gerais - Brazil) 1999
M.S. (Federal University of Minas Gerais - Brazil) 2001

DISSERTATION

Submitted in partial satisfaction of the requirements for the degree of

DOCTOR OF PHILOSOPHY

in

Computer Science

in the

OFFICE OF GRADUATE STUDIES

of the

UNIVERSITY OF CALIFORNIA

DAVIS

Approved:

Shyhtsun Felix Wu, Chair

Matthew Bishop

Frederic T. Chong

Committee in Charge

2010

UMI Number: 3429578

All rights reserved

INFORMATION TO ALL USERS
The quality of this reproduction is dependent upon the quality of the copy submitted.

In the unlikely event that the author did not send a complete manuscript
and there are missing pages, these will be noted. Also, if material had to be removed,
a note will indicate the deletion.

Dissertation Publishing

UMI 3429578
Copyright 2010 by ProQuest LLC.
All rights reserved. This edition of the work is protected against
unauthorized copying under Title 17, United States Code.

ProQuest LLC
789 East Eisenhower Parkway
P.O. Box 1346
Ann Arbor, MI 48106-1346

Copyright © 2010 by
Daniela Alvim Seabra de Oliveira
All rights reserved.

*To Jesus Christ, my Lord and Savior
and to Marcio and Brooke.*

Contents

List of Figures vi
List of Tables vii
Abstract viii
Acknowledgments ix

1 Introduction   **1**
    1.1 Common Defenses and What is Missing 2
    1.2 Dissertation Structure 5

2 VM-Based Full-System Replay for Post-Attack Analysis   **7**
    2.1 About this Chapter 7
    2.2 Introduction 8
    2.3 Nondeterminism 10
       2.3.1 Hardware Interrupts 10
       2.3.2 Input Events 11
    2.4 Log-Based Rollback Recovery 12
    2.5 The Virtual Machine-based Log and Replay Framework 13
       2.5.1 Implementation 13
    2.6 Experimental Results 17
       2.6.1 Linux 17
       2.6.2 Windows XP 18
       2.6.3 Results 18
    2.7 Post-Attack Analysis 20
    2.8 Conclusion 22

3 VM-based Full-System Recovery from Control-Flow Hijacking Attacks   **24**

3.1	About this Chapter	24
3.2	Introduction	25
3.3	Bezoar Design and Implementation	28
	3.3.1 High-level View	28
	3.3.2 Attack Detection	29
	3.3.3 Memory Monitor	31
	3.3.4 Execution Checkpoint, Log and Replay	33
	3.3.5 Design and Implementation	33
	3.3.6 Identifying Network Inputs and Sources	33
	3.3.7 Identifying Malicious Network Inputs	34
	3.3.8 Replaying to Recover from Attacks	34
3.4	Experimental Evaluation	36
	3.4.1 Effectiveness of the Recovery	38
	3.4.2 Performance	41
3.5	Discussion	42
3.6	Conclusions	43

4 Protecting Kernel Code and Data with a Virtualization-Aware Collaborative Operating System — 44

4.1	About this Chapter	44
4.2	Introduction	45
4.3	Attacks on Kernel Integrity	49
4.4	The Integrity Model	50
4.5	High-Level View	51
	4.5.1 Tracking Low Integrity at the VM Level	53
	4.5.2 Tracking Low Integrity Data at the OS Level	53
4.6	Design and Implementation	54
	4.6.1 Writes into Kernel Space	55

		4.6.2	File Integrity	56
		4.6.3	Process Integrity	58
		4.6.4	Module Integrity	58
		4.6.5	Integrity of Dynamically Allocated Kernel Areas	60
		4.6.6	Kernel Threads	61
		4.6.7	Enforcing the Integrity Model	62
	4.7	Experimental Evaluation		63
		4.7.1	Rootkit Attacks	64
		4.7.2	Performance Overhead	67
	4.8	Discussion		68
	4.9	Conclusions		69

5 Related Work — 71

	5.1	Virtual Machines and Security Solutions		71
	5.2	Kernel Integrity Defense		73
		5.2.1	Prevention	73
		5.2.2	Detection	76
	5.3	Taint Tracking		77
	5.4	Deterministic Execution Replay		79
	5.5	Recovery		81
		5.5.1	General Software Failures	81
		5.5.2	Worms and Malware	82
	5.6	Post-Attack Analysis		82

6 Conclusion — 84

	6.1	Summary	84
	6.2	Extensions and Enhancements	87
	6.3	Outlook and Contributions	89

LIST OF FIGURES

2.1	Architectural nondeterministic events.	12
2.2	ExecRecorder VM architecture.	13
2.3	ExecRecorder.	16
2.4	Log size (MB) for different workloads - Linux and Windows.	19
2.5	Log file growth rate.	19
2.6	Performance overhead due to logging - Linux and Windows.	20
2.7	Execution times for wu-ftpd 2.6.0 exploit.	22
3.1	Bezoar.	28
3.2	Bezoar modules.	29
3.3	Defeating Minos information flow tracking.	30
3.4	Memory Monitor - Example 1.	31
3.5	Memory Monitor - Example 2.	32
3.6	Memory Monitor - Example 3.	32
3.7	Identifying the source of network input events.	34
3.8	Discovering the malicious source.	35
3.9	The semi-replay phase.	36
3.10	Performance overhead - SPEC INT benchmarks.	42
4.1	Traditional VM usage model.	45
4.2	Semantic gap.	47
4.3	Integrity model: subjects and objects.	51
4.4	OS-VM communication.	53
4.5	Enforcing the integrity model.	62
4.6	Performance overhead - SPEC CPU benchmarks.	67
4.7	Performance overhead - Unixbench system microbenchmarks.	68

LIST OF TABLES

3.1	Worm exploits.	37
3.2	Instructions between attack injection and detection (in millions).	38
3.3	Number of instructions removed from semi-replay.	40
3.4	Number of client connections x Number of errors during recovery - Average.	41
4.1	Kernel attacks.	63
4.2	Benign modules/devices.	63

Abstract of the Dissertation

Virtual Machine based Mechanisms and Tools for Cyber Attack Prevention, Analysis, and Recovery

Throughout the last decade we have witnessed a widespread use of the Internet and a dramatic change in the way people communicate, do business, and present themselves to the world. It did not take long before criminals started exploring this rich environment seeking fun, pride and later illicit money and even war. In the light of this new generation of malware and attacker's motivations, complete defense strategies must address prevention, detection and response to attacks. In spite of that, the majority of efforts in malware defense currently focus on detection.

This dissertation addresses prevention and post attack analysis and recovery in the context of virtual machine (VM) environments. It provides a study of full system replay for post-attacks analysis where the execution of an entire system from a checkpoint can be faithfully replayed with low performance/space overhead. Building on this research, it describes the application of this replay approach on post-attack recovery from control-flow hijacking Internet worms. Finally, this dissertation challenges the traditional VM usage model that advocates placing security mechanisms only in the VM layer, letting the guest operating system (OS) run unaware of virtualization. It shows how collaboration between the guest OS and a VM helps bridge the semantic gap between these layers and provides stronger system protection. The dissertation additionally reports on implementations and proof-of-concept prototypes of these mechanisms, showing them to be effective for their respective scope. The implementations and prototypes validate our proposed approaches and have no false positives or negatives (in the context of prevention and for all the attacks used in our experiments), low performance/space overhead (post-attack analysis) and address zero-day attacks (post attack recovery).

ACKNOWLEDGMENTS

It is extremely difficult to express gratitude in a few paragraphs to all that provided me support along these six years in the PhD program. First and foremost I thank God: "*All the glory, all the victory, I give You, because without You, I would not be here*". I also thank God for leading my way through all the joyous and difficult moments during these years, and for all the wonderful people at UC Davis that positively changed my life for the better. God provided me with everything I needed: emotional support, an excellent advisor, a research topic, papers and a job after completing my degree.

I would like to express extreme gratitude to Prof. Felix Wu. He believed in me since the beginning, supported me through all these years, and invested a great amount of his time in me and on my research topic. Most importantly, he represents to me a role model of a talented researcher and a great human being. He taught me not only how to be a great researcher, but also how to be an excellent advisor and a teacher. I could never have thought of a better advisor. I also would like to thank Prof. Fred Chong and Zhendong Su for all the support, advice and critical feedback they provided me through these years at UC Davis, especially in the first couple of years of my research work. I thank Prof. Matt Bishop and, again, Prof. Fred Chong for taking the time to participate in my committee. I also thank Prof. Hao Chen and Jed Crandall for participating in my qualifying exam committee. Their feedback has definitely improved the quality of my work.

I also thank the CS department administrative and technical staff, especially Kim Reinking, Jessica Stoller and Babak Moghadam.

I thank, again, Jed Crandall for the invaluable help he gave through these years. Our research discussions, his feedback and, most importantly, the example he set as a creative, passionate and hard-working researcher positively impacted my research and my future career. I also thank all my friends and colleagues from the Security lab and the CS Department (including those who already graduated), especially Shaozhi Shawn Ye, Sean Peisert, Xiaoming Lu, Matt Spear, Juan Lang, Gary Wassermann and Earl Barr. They gave me feedback and advice on many situations and on paper drafts and practice talks and this really improved the quality of my research significantly.

I thank the Brazilian people and government for the excellent and free education I was provided with during my undergraduate studies and MS program at the Federal University of Minas Gerais (UFMG).

Finally I would like to thank those that provided me with the emotional support and stability I needed during this journey. First I thank my husband Marcio. His constant love, friendship, stability, patience and support has made my life so much better! He was by my side through difficult moments, tough decisions and, of course, during good time as well. Without him, everything would be much more difficult and less fun. *Eu te amo muito!* I thank my family for supporting me through these years and helping me when I had my daughter during the PhD studies, especially my mom Gesa, my dad Enio, my mother-in-law Maria Luiza, my father-in-law Jose Agostinho and my sister Camilla. I thank my Brazilian friends whose company through e-mail and online chats made my life here so much easier: Joao, Marcelo, Valdo, Lorena, my girlfriends from UFMG (Junia, Pat, Gisele, Fabiana and Pati), Benicio, Melissa, Fabiana, Fernando, Claudine and Ana. I thank my Brazilian friends in California who made me feel at home here in the USA: Reuber, Juliana, Edson, Marilene, Liliana, Alesio, Nadia, Bia, Desiree, Guilherme, Patricia and Sueli. Also, I could never forget my beloved German shepherd Duke. For countless times he relieved my stress just by looking at me with his loving light brown eyes and hugging me with his giant paws and wagging tail. And I thank the most precious thing I have in this life: my daughter Brooke. *Minha Ilarinha*, you have no idea how much I love you!

Chapter 1

Introduction

A decade ago terms like malware, remote attacks, threats, cyber crime and cyber war sounded slightly farfetched. At that time a small percentage of the world's population had Internet access and the majority of those accesses were through dial-up. Throughout the last decade we have witnessed not only a widespread use of the Internet but also a dramatic change in the way people communicate, do business, do banking, make friends and present themselves to the world. The number of Internet users has also reached a quarter of the world's population (1.7 billion) [1]. In this new and rich environment we have millions of people from all sorts of origin, age, gender, interests and technical knowledge using software/hardware that were not developed taking security into account. As a result, it did not take long before hackers, criminals, organized groups, terrorists and nation states started exploring this environment, seeking fun, pride, vandalism (during the first half of the last decade) and later illicit money, terrorism, sabotage and even war.

To make matters worse, recent security reports emphasize that this state of affairs is expected to worsen in the next few years [1–3]. In the light of this new situation, how should the security research community, enterprises and nation-states prepare themselves for the next decade? In a recent conference panel, the NITRD Senior steering group on Cybersecurity R&D advocated a *"changing the game"* approach against malware and attacks [1], *i.e.*, if we cannot directly win this arms race against attackers, we need to change the game. Their recommendations advocated for diversity on defense solutions against the next generation of malware and attacks we are going to face during this new decade. For example, morphing the board and changing the terrain with non-persistent virtual machines and adaptive networks and raising the stakes by increasing the risk and cost for attacks. Thus, complete defense strategies must encompass techniques and tools addressing prevention, detection and response to malware and attacks.

1.1 Common Defenses and What is Missing

Currently, the majority of efforts in malware defense are in the realm of detection, *i.e.*, recognizing that an intrusion, infection or attack is taking place and possibly stopping it afterwards. Detection efforts are mostly represented by intrusion detection systems and anti-virus software.

For instance, defense solutions against Internet worms have been addressed through detection approaches and attack signatures. Worms are a subclass of viruses whose primary vector is the network, *i.e.*, they get transferred to a victim host through data coming from the network such as an e-mail attachment or network protocol messages [5, 6]. A worm exploit is an instance of code or bytes conforming to a certain application protocol that explores a vulnerability in a networked application so that management information is corrupted with malicious data. When this data is used by the architecture, the OS or the vulnerable program, the malware is activated and allows an attacker to take control of the system [7]. This is possible because, in general, management information is stored adjacent with or in-band to regular data (*e.g.*, a string variable). Detection-based defense approaches employed against worms usually involve information flow tracking, address-space randomization and memory protection methods. Usually, after attack detection, the system crashes or reboots, or the offending application is terminated, which causes disruption of service for many users. Even if the system continues its execution after the attack (by killing the offending application, for instance), it could not be able to proceed.

Another example of common malware whose defense strategies are usually proposed through detection are rootkits, which are a type of stealthy malware that enable attackers who have gained administrator privileges or access to the system (usually through vulnerability exploitation or social engineering) to hide or covertly perform malicious actions and maintain control of the system. They have been increasingly used by attackers bundled with other type of malware such as Trojans, droppers, bots and key loggers. Kernel rootkits can corrupt, change, influence and add malicious behavior to the OS kernel, which compromises the integrity of the entire system. The solutions for kernel integrity defense presented in the literature focus mainly on detecting changes in kernel control flow or code, and most of them are not effective for detecting attacks that involve value manipulation of kernel data structures.

Although detection schemes are essential for malware defense, they, alone, are inefficient for countering this new generation of attacks and malware. We also need to employ effective prevention and response mechanisms to better secure a system against compromise.

We need to employ prevention mechanisms because the current generation of threats spreads slower than their predecessors to avoid detection [1]. Also, when malware is detected it is usually not alone: the host is already infected with several unrelated malicious software and rootkits. Moreover, the consequences of an attack can last much longer after the immediate damage is done. We also need to employ appropriate measures of response and recovery. After an attack what can be done to allow a system to continue providing its services? For instance, many servers do not tolerate long downtimes caused by OS rebooting and reinstallation, file disinfection and recovery from backup. Also, in many countries critical utilities such as water and the electrical grid are controlled via networked computer systems. Manual recovery procedures, performed by systems administrators after an attack, are slow and error-prone and may not leave the system in a stable state, it may be left at a corrupted state after the attack or may be still infected by other malware.

And finally, we need collaboration between layers of hardware and software so that we can bridge the semantic gap between them. Although there are some APIs among these layers, the current interface is not as strong and as collaborative as needed. The next generation of security solutions will need to access and manipulate information at several levels of abstraction to be effective.

This dissertation addresses cyber attack prevention and post attack analysis and recovery in the context of virtual machine environments. It also proposes collaboration between the OS and the architecture layer as an effective approach to build stronger security solutions. It contributes with mechanisms and tools for preventing compromise, analyzing attacks offline with low overhead, and allowing a system to continue execution correctly after an attack in most of cases. It additionally reports on implementations and proof-of-concept prototypes of these mechanisms, showing them to be effective for their respective scope. The implementations and prototypes validate the proposed approaches and additionally, have no false positives or negatives (in the context of prevention for all the attacks used in our experiments), low performance/space overhead (post-attack analysis) and address zero-day attacks (post-attack recovery).

1.2 Dissertation Structure

First, in Chapter 2 this dissertation discusses a full system replay approach for offline post-attack analysis, where the execution of an entire system can be faithfully replayed from a checkpoint with low performance/space overhead. This chapter also provides a characterization of nondeterminism in terms of architecture level events, which are the minimum amount of information required to be logged so that the system execution can be reproduced. Its main advantage compared to previous approaches is that it works at the architecture level, which gives us fine-grained control of the replayed events.

Chapter 3 presents the application of this replay approach analysis on post-attack recovery against memory-injection attacks, especially Internet worms. The main issue was to maintain the availability of a host after an attack, where much of the system state could have gone corrupted. Basically, we tracked down the source of network bytes in the system and, after an attack, we replayed the checkpointed run while ignoring malicious inputs. The challenge of this work was replaying the execution while removing malicious events. A couple of other groups have worked on something similar but they really did not address the issue of shared corrupted state, which we did. In this work we faced some challenges related to how to reproduce system entropy (that is associated with the number and type of nondeterministic events in the system) while removing malicious nondeterministic events from the run. Our challenge in controlling entropy was due to the fact that it is an OS concept, while our recovery mechanism was designed to operate at the architecture level. This prompted us to believe that if we could add collaboration between the guest OS and the architecture layer (represented by a VM) we could build stronger, more flexible and fine-grained security solutions.

Chapter 4 continues showing how collaboration between an OS and a VM layer can help us bridge the semantic gap between these layers which is the root of several shortcomings in many current security solutions. This chapter presents a model of a collaborative architecture between a virtualization-aware guest OS and a VM. This model challenges the traditional VM usage model, which advocates placing all the security mechanisms inside a trusted VM and letting the guest OS run unaware of virtualization. The chapter also describes a case study where we employed this collaborative architecture to prevent tampering against the OS kernel code and data.

Chapter 5 discusses related work and finally, Chapter 6 brings our conclusions, main contributions and future work.

Chapter 2

VM-Based Full-System Replay for Post-Attack Analysis

2.1 About this Chapter

Log-based recovery and replay systems are important for system reliability, debugging and postmortem analysis/recovery of malware attacks. These systems must incur low space and performance overhead, provide full-system replay capabilities, and be resilient against attacks. Previous approaches fail to meet these requirements: they replay only a single process, or require changes in the host and guest OS, or do not have a fully-implemented replay component. This chapter studies full-system replay for uniprocessors by logging and replaying architectural events. To limit the amount of logged information, we identify architectural nondeterministic events, and encode them compactly. Here we present ExecRecorder [8], a full-system, VM-based, log and replay framework for post-attack analysis and recovery. ExecRecorder can replay the execution of an entire system by checkpointing the system state and logging architectural nondeterministic events, and imposes low performance overhead (less than 1% on average). In our evaluation its log files grow at about 5.1 GB/hour (arithmetic mean). Thus it is practical to log on the order of hours or days between checkpoints. It can also be integrated naturally with an IDS and a post-attack analysis tool for intrusion analysis and recovery.

2.2 Introduction

Log-based recovery[9–15] and replay [16–23] systems are important for system reliability and security. Recent work has also used a replayer to perform postmortem analysis of malware attacks [16]. After an attack, one may replay the attack sequence using off-line analyzers. Replay systems can also be used for recovery from malware attacks by integrating them with an intrusion detection system (IDS) and an analysis tool. Upon detecting an attack (with an IDS) and discovering the execution point at which the attack happened (with an analysis tool), one can rollback the system execution to an earlier checkpoint and disable particular effects of the attack.

Replay and recovery systems are generally based on three components: checkpoint, log and replay. The checkpoint component captures a snapshot of the current state of a system at specific times. The log component records the nondeterministic events that affected system execution since the checkpoint was taken. The replay component uses the information logged along with the checkpoint to deterministically replay the system execution during that specific run (the sequence of states a system passes through during execution, represented by the partial ordering of events sent and received, and also local events [10, 14]).

These systems must incur low space and performance overhead, provide full-system replay capabilities, and be resilient against attacks. Previous approaches fail to meet these requirements. Most replay only a single process or application [17–20]. Those that address the whole system require changes in the host and guest OS [16], or do not have yet a fully-implemented replay component [23].

This chapter studies full-system log-and-replay for uniprocessors by logging and replaying architectural events. In order to limit the amount of information that needs to be logged, we characterize architectural nondeterminism, both by identifying nondeterministic events and by encoding them compactly. Working at the level of architectural events enables a full-system replay that is flexible with respect to the OS and the applications.

We have implemented our system as ExecRecorder, a full-system, VM-based log and replay framework for post-attack analysis and recovery. It can replay the execution of an entire system (not only a process or a distributed application in isolation) by checkpointing the complete system state (virtual memory and CPU registers, virtual hard disk and memory of all virtual external devices) and logging all architectural nondeterministic events. The checkpoints can be taken at any time and the replay does not need to start from a powered-off machine. Our strategy for checkpointing the hard disk (HD) is efficient (based on copy-on-write), and is achieved by using committable/rollbackable disk images. Its main advantage compared to previous approaches is that it works at the architectural level, which allows fine-grained control of the replayed events and flexible post-attack analysis.

ExecRecorder runs as part of a VM, is not easily accessible to malware, yet still gives a detailed view of the system execution. In order to perform post-attack analysis and recovery, there should be integration and cooperation among the IDS, analysis tool and replay system.

ExecRecorder imposes low performance overhead (beyond that of the VM). In our evaluation, which includes multiple workloads in both Windows and Linux, its log files grow at about 5.4 GB/hour (arithmetic mean). Thus it is practical to log on the order of hours or days between checkpoints. It can also be integrated naturally with an IDS, such as Minos [24], to determine when an exploit has occurred. An analysis tool, such as DACODA [25], can point where the exploit was detected in order to understand the exploited vulnerability.

The rest of this chapter is organized as follows. Section 2.3 characterizes architectural nondeterministic events and povides a compact format encoding for them. Section 2.4 discusses log-based rollback recovery. Then section 2.5 presents ExecRecorder and details its implementation, followed by section 2.6, which presents our experimental evaluation. In section 2.7 we describe an example of post-attack analysis using ExecRecorder, and conclude in section 2.8.

Our contributions are as follows. First, we provide a characterization of architectural nondeterminism for uniprocessors and a compact format encoding. Second, we present a low-overhead, full-system log and replay framework for uniprocessors that runs integrated with a VM and does not require any modification in the guest or host OS. Third, we show with a practical example how such system can streamline the analysis of a zero-day exploit.

2.3 Nondeterminism

Models of real-world systems necessarily exclude some details of the systems they model. Within a model, a nondeterministic event is one that causes a state transition that is not fully determined by the previous state, *i.e.*, the event could not have been predicted with certainty from knowledge only of the model's previous state(s). Nondeterministic events are the minimum amount of information necessary to reproduce the system execution. Deterministic events, on the other hand, can be regenerated if a system executes the same set of instructions starting from the same memory state (checkpoint). For systems based on uniprocessors, the nondeterministic architectural events are hardware interrupts and input events.

2.3.1 Hardware Interrupts

External devices such as the hard disk (HD) or the keyboard generate interrupts asynchronously with regard to the processor clock [26]. Although it is possible to model the behavior of a certain device with great accuracy, the exact time at which an interrupt is generated is unpredictable.

We encode an interrupt as an integer representing the difference between the tick of the last event and the tick at which the interrupt occurred, and an integer encoding the interrupt value, *i.e.*, the IRQ line number. In our VM, the tick means the number of instructions executed before that event happened [27]. This definition assumes implicitly that the time for each instruction's execution is the same. Logging just the tick difference allows us to considerably reduce the amount of logged data. For example, if the tick at which an interrupt occurred is 4294967298 and the tick of the last event is 4294967297 (both requiring 8 bytes for recording), we log just 1 as the tick information for the interrupt event. This approach decreases the number of bytes required to record the event's timing from 8 bytes to a maximum of 4 bytes (we have observed that the tick difference does not exceed 2^{32})

In this study we have considered the following external devices: PIT, CMOS, HD, keyboard, mouse, network card and serial and parallel port devices. PIT interrupts are generally regarded as deterministic events. We are characterizing them as nondeterministic because, although most of its interrupts occur at a fixed frequency, there is some nondeterminism coming from noise in the device's crystal-controlled oscillator and from our VM's PIT implementation. As a result, these interrupts are not entirely predictable.

2.3.2 Input Events

These are events where data from a device is read to main memory or a register where the CPU can process it. The exact time at which these data become available to the CPU is also unpredictable, even though the behavior of external devices can be modeled. Moreover, the bytes coming from such events are also nondeterministic because they can come from user input, network or the environment (for example,

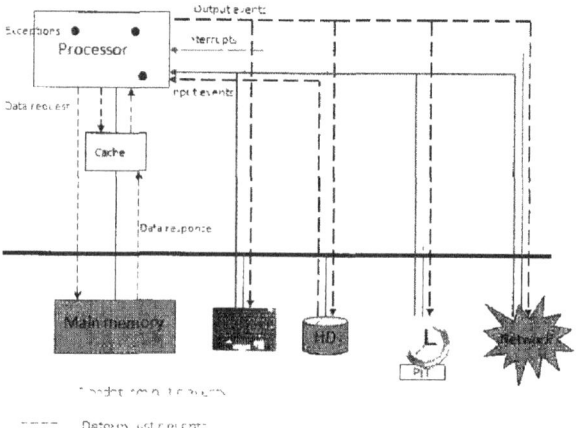

Figure 2.1. Architectural nondeterministic events.

the current time). An important exception are the bytes coming from the HD. The value of these bytes is based solely on the contents of the disk and the disk requests made by a program. Thus, we characterize input events coming from the HD as deterministic. We encode an input event as an integer representing the tick difference and an integer representing the byte(s) read. Figure 2.1 illustrates our characterization of architecture level nondeterminism.

2.4 Log-Based Rollback Recovery

Log-based rollback recovery is a technique used to achieve fault-tolerance in distributed systems and also to allow the replay of an application or system for debugging or post-attack analysis. The main idea is to combine checkpoints with a log file. The checkpoint is a snapshot of the system state. The log file contains enough information to reproduce all nondeterministic events that occurred during a run. All messages received by the system during a run are classified as nondeterministic. Log-based rollback recovery relies on the *piecewise deterministic* assumption [9, 10], which states that all nondeterministic events that a process executes can be identified and all information necessary to reproduce these events can be logged.

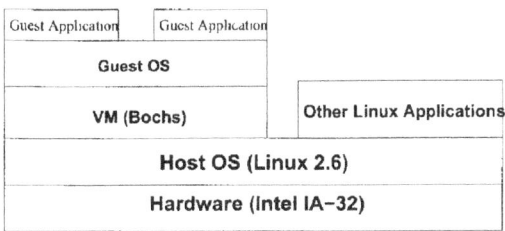

Figure 2.2. ExecRecorder VM architecture.

The set of information needed to reproduce a nondeterministic event is called the *determinant* of the event [11].

Besides the checkpoint and the log components, there is also a replay component. The log component continuously records determinants of nondeterministic events in a non-volatile medium. The replay component recovers the system state captured by a checkpoint (usually the latest one) and uses the determinants in the log file to reproduce the execution of a run. The amount of data recorded plays an important role on the performance of such systems. It will directly impact the overhead incurred by the log and replay components.

2.5 The Virtual Machine-based Log and Replay Framework

2.5.1 Implementation

We have used Bochs [27] as the VM of our experiments. It is a hosted-VM (installed on top of a host OS) that emulates the IA-32 Pentium architecture. The host OS is Linux 2.6. Our VM is currently integrated with Minos [21], a microarchitecture to catch attacks, and DACODA [25], a post-attack analysis tool. We have designed this system to run as a honeypot. Figure 2.2 illustrates this architecture. ExecRecorder has three main components (checkpoint, log and replay) and was implemented as an extension to Bochs.

2.5.1.1 Checkpoint

This module is executed immediately before the logging of a system run. It is responsible for saving the system state (virtual main memory, CPU registers, HD and memory from external devices) at the current instant of time. We have implemented it by duplicating the Bochs VM process via the *fork* system call. After the fork, the parent process waits in the background for a SIGUSR1 signal while the child process continues its execution. The suspended parent represents the frozen state of the system at the time the checkpoint was taken.

The checkpoint of the virtual HD is achieved by using Bochs undoable disk mode [27]. An undoable disk is a committable/rollbackable disk image. It is based on a read-only disk image combined with a file (*redolog*) that contains all changes made to the read-only image. After a run the *redolog* file can be merged to the read-only image or simply discarded. ExecRecorder always starts the VM with the read-only disk image. When a checkpoint is taken, the child process continues its execution with a new *redolog* file, which is initialized with the contents of the parent process *redolog* file.

2.5.1.2 Log

The log component records in the host HD enough information about the nondeterministic events happening in the system so that they can be later replayed.

In order to correctly replay input events we need to log more information than just the characterization of nondeterministic input events given in section 2.3.2. Although an input event can be solely characterized by the time or tick at which it occurred and the bytes read, our replay component needs enough information about the input instruction itself to correctly reproduce it. For example, for the Intel IA-32 architecture [28] we have input instructions to transfer a (or a string of) byte(s), word(s), or a double word(s) between an I/O port and a CPU register.

To simulate an I/O port instruction, our VM implementation requires knowledge of the number of bytes being transferred and whether it is to a register or memory. In theory, this type of information is not required in the log file.

Although we have considered all input events from the HD as deterministic, the replay component still needs information about the HD input instruction to reproduce it. The reason is that, during replay, we need to make the disk requests synchronized with the tick at which an HD interrupt occurs. If we do not log and replay such instructions, it may occur that (at least in our Bochs VM) a HD interrupt is raised before the intended bytes are read from the disk. Note that we do not log the bytes read from the disk but only information about the input instruction. FDR [23], on the other hand, logs all values returned from I/O loads.

The format of our log files is as follows:

- Event type (1 byte): input event, interrupt or the handling of an interrupt by the CPU (the last one is Bochs-specific);

- Tick difference (1 bytes);

- Input events-specific: port address (2 bytes), bytes (1 bytes, not logged for HD), flags (1 byte encoding information about number of bytes being transferred and whether it is to a register or memory), memory address (4 bytes, logged only if bytes are being transferred to memory);

- Interrupts-specific: IRQ number (1 byte).

2.5.1.3 Replay

After the logging of a run this module can be called to reproduce the system execution from a certain checkpoint. The child process wakes up the parent process with a SIGUSR1 signal. The parent process, which captures the system state at the point the checkpoint was taken, resumes its execution. The virtual disk image used is also the one at the time the checkpoint was taken. However, all interrupts

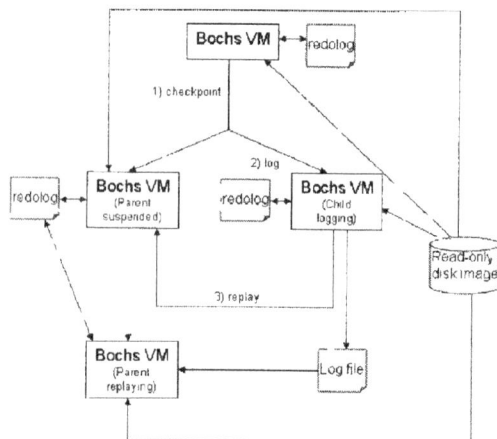

Figure 2.3. ExecRecorder.

or input events that may be generated are disabled and do not affect the state of the system. Being in replay mode, the VM uses all information recorded in the log file to reproduce the events at the tick at which they happened during the log phase. Figure 2.3 illustrates how ExecRecorder works.

The log and replay framework also acted as an oracle in validating our characterization of architectural nondeterministic events. We have validated our characterization of nondeterminism by trying to remove each type of nondeterministic event and then replaying the system execution. The exclusion of a certain type of nondeterministic event would prevent a successful replay.

2.5.1.4 Multiprocessors and DMA Discussion

Our proof-of-concept implementation currently does not address multiprocessors and DMA. However, our approach can be extended, in principle, to include them. A first direction for this future work is to extend our VM to model the cache subsystem, DMA and the bus. From these models we can extend ExecRecorder to log DMA writes and the minimal subset of memory races according to the algorithm proposed in the FDR design [23].

2.6 Experimental Results

In our experiments we have analyzed, for Linux and Windows, the size of the log files generated by ExecRecorder and the log files growth rate when we varied the workload in the guest OS. We have also analyzed the performance overhead incurred by the logging component. We have studied the system in the following situations: running a Web server, which is receiving a burst of requests in a noisy campus network, executing intensively its CPU and disk, running multitask activities, and idle.

We have selected a set of publicly available applications as our workloads. For each one of our experiments we ran each workload three times and averaged the results obtained. The workloads chosen for Linux and Windows were independent from one another because we have used publicly available workloads or benchmarks and, in general, they were developed for a specific OS. The experiments were executed on a Pentium 4 SMP with 2 3.2 GHz CPU's and 1 GB of RAM.

2.6.1 Linux

We have selected two workloads for our Linux 2.4.21 guest OS. The first one tests the system running the Apache Web server [29]. It generates, from an external network, a burst of 2000 requests to fetch a 3K html document.

The second workload was UnixBench [30], which is a benchmark suite for Linux that integrates CPU, file I/O, process spawning and other workloads. The following tests were performed: Dhrystone 2 using register variables, arithmetic, system call overhead, pipe throughput, pipe-based context switching, process creation, exec throughput, file system throughput, concurrent shell scripts, compiler throughput, and recursion.

2.6.2 Windows XP

We have selected three workloads for Windows. The first is the same used to test Linux as a Web server [29]. We tested the Apache Web server in Windows by generating 200 requests to fetch a 3K html document. The requests were also generated from an external network, where, in this case, we have generated one request per second. We have inserted a light load in our Web server for Windows, because it could not handle well more than 200 HTTP requests per second.

The second workload was Microsoft SQLIO [31], a disk subsystem benchmark tool. It generates disk workload so as to simulate aspects of the I/O workload of the Microsoft SQL Server. In our tests, we had one thread reading for approximately two minutes from a file using 2 KB IO's over 128 KB stripes with 64 IO's per run.

The third workload was an implementation of the Sieve of Eratosthenes. Our goal was to generate a CPU-intensive workload. We have chosen to use this algorithm not only because it is usually part of several well-known CPU benchmarks, but also because publicly available CPU benchmarks for Windows were interactive and we did not want the user response time to influence our results.

2.6.3 Results

Figure 2.4 shows how the size of our log files varied for each considered workload for Linux and Windows, and Figure 2.5 presents the corresponding log file growth rate in GB/hour.

Although our choice of applications does not represent a characterization of a certain type of workload, we observe that I/O-intensive applications, especially those that extensively use the HD, tend to have a larger log file growth rate. Also, our results show that ExecRecorder is feasible and practical for different types of workloads provided that the frequency at which checkpoints are taken is chosen appropriately, considering the amount of disk space available for logging. As the cost of HDs is relatively low, checkpoints can be taken every hour, twice a day or

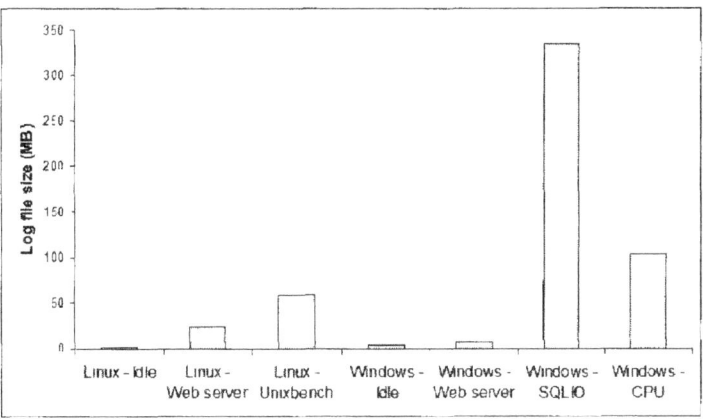

Figure 2.4. Log size (MB) for different workloads - Linux and Windows.

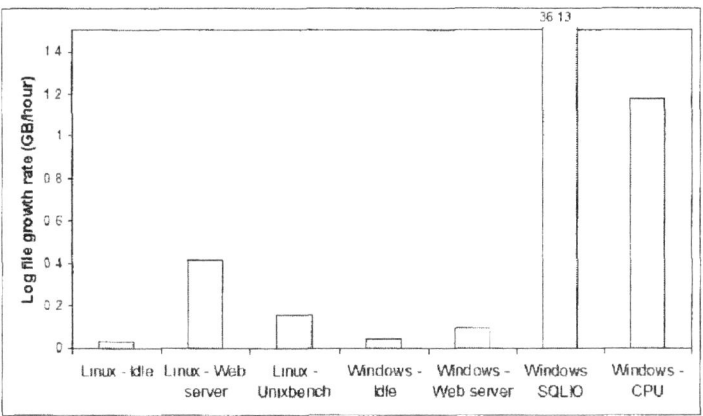

Figure 2.5. Log file growth rate.

every day, depending on the demand of the application. Although the redolog file (section 2.5.1.1) should count as part of the non-volatile storage necessary, we did not consider it as part of the ever-growing log. This is because the redolog file can be at most the size of the original HD no matter how long we run a benchmark.

Figure 2.6 illustrates the performance overhead of the logging component for our selected workloads. For all cases the overhead due to logging is low (less than

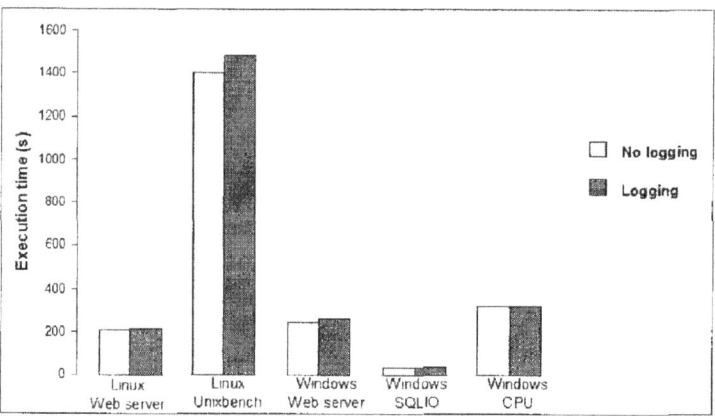

Figure 2.6. Performance overhead due to logging - Linux and Windows.

1% on average). We have not shown performance results during replay because according to Elnozahy and Alvisi [9], it has been observed that in replay mode the system can run considerably faster than in normal execution. During normal execution a process may block waiting for I/O events, while during replay all events can be immediately replayed.

2.7 Post-Attack Analysis

Here we describe a practical example of using ExecRecorder to perform post-attack analysis. In this experiment we have Minos [24] as our IDS and DACODA [25] as our analysis tool, according to the architecture shown in Figure 2.2.

Minos is a security-enhanced microarchitecture that prevents attacks that hijack program control flow, for example, buffer and integer overflows, format string exploits, double free calls, heap corruption and register spring attacks. Every 32-bit word of memory and every 32-bit general purpose register in the Intel x86 architecture is augmented with one tag bit which represents the integrity level of this word (zero meaning low integrity and one high integrity). This bit is set by the kernel based on the trust it has for the data. The basic assumption is that any control transfer (instructions such as jump, call, and return) involving untrusted data is a system vulnerability and a hardware exception traps to the kernel whenever this occurs.

DACODA is a tool that analyzes attacks using symbolic execution. It labels each byte coming from the network with a unique identifier and tracks these bytes in the system during their lifetime. When Minos catches an attack, DACODA provides information about it, such as processes involved, if the attack involved kernel or user processes, tokens that compose the attack trace and the predicates found. A limitation of this tool is the performance overhead it incurs, because for each instruction executed, DACODA has to perform symbolic execution. This overhead is exacerbated for exploits that require considerable amount of computation such as Code Red II and the ASN.1 exploit. IntroVirt [32] also uses predicates to detect intrusions. The difference between the two is their goals. IntroVirt checks if a system has been exploited in the period between vulnerability discovery and patch release, while DACODA analyzes and generates signatures for zero-day exploits.

ExecRecorder, Minos and DACODA currently run as extensions to Bochs. To integrate ExecRecorder with Minos and DACODA we just need to log and process more information. For Minos we have to log the integrity bit of every word transferred in input events and for DACODA we have to log all incoming network packets because their bytes need to be labeled in the order that they were received by the network card and not in the order delivered to the CPU (the bytes are usually reordered in the network card).

A solution is to turn off DACODA in our honeypot and only execute it offline using ExecRecorder. Our honeypot executes Minos along with ExecRecorder in log mode, which incurs very low performance overhead. When Minos catches an attack, we use the log file generated since the last checkpoint and analyze the attack off-line with DACODA.

Here we analyze, for an exploit of the wu-ftpd 2.6.0 vulnerability [33], the size of

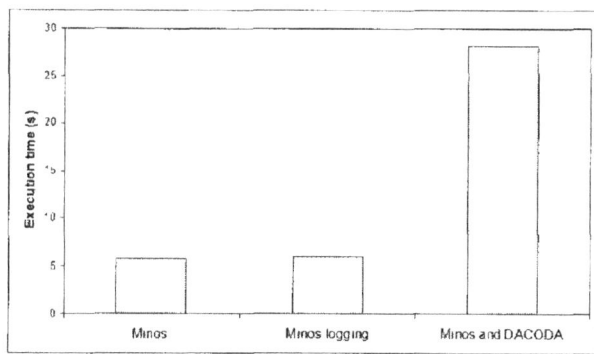

Figure 2.7. Execution times for wu-ftpd 2.6.0 exploit.

the log file generated and the execution time of the attack in three situations: when it executes with only Minos on, when it executes with Minos and ExecRecorder in log mode, and when it executes with Minos and DACODA on. Figure 2.7 shows the exploit execution time for these three situations. The log file for the exploit is 1.769 MB.

DACODA provided us with the following information about this attack: (1) it has a total of 2888 predicates and all of them were found in user space, (2) the process involved is wu-ftpd, (3) the longest signature token has 283 bytes, and (4) the token length histogram as "Number(size in bytes)" is 4(283), 1(119), 4(11), 1(10), 1(9), 1(6), 4(5), 3(4), 4(3), 10(2), 41(1).

2.8 Conclusion

In this chapter we presented ExecRecorder, a full-system, VM-based log and replay framework for uniprocessors to perform post-attack analysis. It addresses the limitations found in current replay systems by providing full-system replay capabilities and low performance overhead without requiring any OS changes.

The lessons we have learned can be summarized as follows. We can considerably decrease the amount of logged data by recording the tick or time difference of an event and the last one, instead of its absolute timing value. Also, although an input event can be characterized by its timing and bytes only, a replayer usually will need information about the input instruction itself to correctly reproduce the event. This extra information will vary depending on the system architecture or the VM implementation. Input events from the HD are deterministic but we still need to log information about their associated input instruction to synchronize the event with its corresponding interrupt. The HD bytes, however, do not need to be logged

ExecRecorder approach is promising and it was recently employed to analyze covert channels through repeated replays with varying confidential data [34]. In Chapter 3 we describe how we employed full-system replay for post-attack recovery against control-flow hijacking attacks. In Chapter 6 we discuss future work associated to our replay approach.

Chapter 3

VM-based Full-System Recovery from Control-Flow Hijacking Attacks

3.1 About this Chapter

System availability is difficult for systems to maintain in the face of Internet worms. Large systems have vulnerabilities, and if a system attempts to continue operation after an attack, it may not behave properly. Traditional mechanisms for detecting attacks disrupt service, and current recovery approaches are application-based and cannot guarantee recovery in the face of exploits that corrupt the kernel, involve multiple processes or target multithreaded network services. This chapter presents Bezoar [35], an automated full-system virtual machine-based approach to recover from zero-day control-flow hijacking attacks. Bezoar tracks down the source of network bytes in the system and after an attack, replays the checkpointed run while ignoring inputs from the malicious source. We evaluated our proof-of-concept prototype on six notorious exploits for Linux and Windows. In all cases, it recovered the full system state and resumed execution. Bezoar incurs low overhead to the virtual machine: less than 1% for the recovery and log components and approximately 1.4X for the memory monitor component that tracks down network bytes, for five SPEC INT 2000 benchmarks.

3.2 Introduction

Control-flow hijacking Internet worms perform their attacks by overwriting control data in a victim host, which allows them to perform arbitrary malicious actions. Software systems cannot be guaranteed to be free from vulnerabilities because designers and programmers do make mistakes and current verification and testing techniques cannot assure that a piece of software meets its specification in the presence of errors or bad inputs. Thus, techniques for software fault tolerance are necessary to guarantee availability in real-world systems.

Fault-tolerance is comprised of four phases [36]: (i) error detection, (ii) damage confinement and assessment, (iii) error recovery, and (iv) fault treatment and continued system service. In the context of control-flow hijacking attacks, intrusion defense systems usually address (i) by detecting and stopping the exploit using information flow tracking [24, 37–39], address-space randomization [10, 41], memory protection [42, 43] and other methods [44]. Usually, after attack detection, the system crashes or reboots, or the offending application is terminated, which causes disruption of service for many users. For example, if an IDS (intrusion detection system) reboots the system, it will drop non-malicious clients, lose recently acquired data, and experience down-time as it restarts and replenishes cached data for services like DNS. Even if the system continues its execution after the attack (by killing the offending application, for instance), it could not be able to proceed. The exploit could have corrupted areas of memory used by non-malicious processes and the system could eventually crash or hang. Also, the attacker could have damaged critical system files or data structures and the continuation of the host execution could lead to a situation of error or inconsistency.

Some defense systems perform damage confinement and assessment in varying degrees by analyzing an attack once it is stopped and patching the vulnerability with filters or signatures [25, 45–49]. More recently the research literature proposed some techniques to address post-attack recovery [50–55] and most of them are application-based [50, 52–55], i.e., they aim at recovering a specific process (usually the victim one) after a worm attack.

Application-based recovery approaches cannot guarantee recovery from the new generation of attacks. As explained in [25, 56], kernel exploits are expected to be common in the near future. An exploit does not necessarily need to involve an user-space process because a vulnerability in the kernel will allow the execution of malicious code in supervisory mode. Further, certain OSes, such as Windows, handle network traffic in the kernel and certain exploits, such as heap overflows [57], modify values within the heap management information structure and corrupt memory outside the victim process address space. Also an exploit may involve multiple processes and use multiple stages (e.g. mnd exploit [58]), and certain network services are multithreaded. In a multithreaded Web server, for instance, the malicious network packet can be first received by a thread listening on a certain port, but the memory corruption itself may only occur in another thread, spawned just to treat the newly arrived TCP connection. For an application-based approach it may not be trivial to decide which thread needs to be recovered: the one that first received the malicious network input or the one that actually caused memory corruption? Other limitations of current recovery approaches include: not recovering damage to the file system; losing state related to interprocess communication, messages, signals, and resources (e.g., files opened, processes spawned and pages allocated); undoing the effects of the attack speculatively [52, 53, 55] and requiring the application source code [54, 55].

This chapter presents an automated full-system virtual machine-based approach to recover from zero-day control-flow hijacking attacks, which we call Bezoar[1]. It incurs low overhead to the virtual machine and is OS and application independent. Bezoar combines and integrates the following techniques: lightweight full-system checkpoint (based on copy-on-write), log/replay, and tracking of the propagation of network data into the memory.

Figure 3.1 provides an overview of our recovery strategy. The main idea is that during normal execution the system takes periodic checkpoints while logging all architectural nondeterministic events: interrupts and input events as described in Chapter 2(1). The system also tracks down how data from the network propagates into the memory (including registers and memory from external devices). Each memory byte will be associated with zero, or one, or more network sources. In the event of an attack (2), we discover its source (3), go back to the most recent checkpoint (4) and replay the system execution while ignoring all network packets coming from the malicious source (5). After the replay with recovery phase, the system can continue its execution normally (6). We have implemented this approach as a proof-of-concept prototype using the Bochs virtual machine [27].

We have successfully tested our approach using six notorious exploits (Code Red II, Blaster, Slammer, *innd*, *rpc.statd* and *wu-ftpd*) in Windows and Linux. Our log and recovery components incur very low overhead to the VM (less than 1%). The component that tracks down network bytes into memory incurs a slowdown of approximately 1.1X.

The rest of the chapter is organized as follows. Section 3.3 describes in detail Bezoar's design and implementation. In section 3.4 we present the experiments we have performed to validate our recovery approach and in section 3.5 we discuss Bezoar issues and limitations. Our conclusions are presented in section 3.6.

---

[1] Bezoar is a stone taken from the stomach of ruminant animals that was formerly believed to be an antidote from most poisons

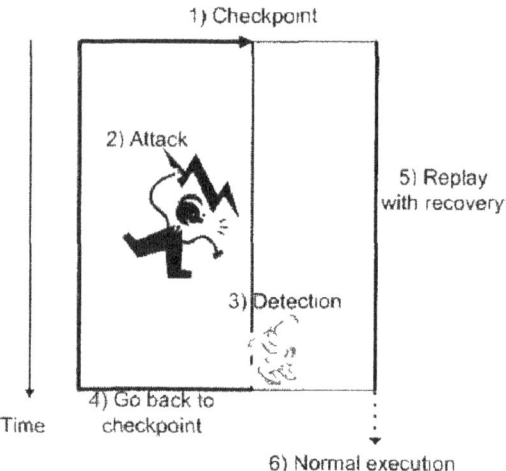
Figure 3.1. Bezoar.

3.3 Bezoar Design and Implementation
3.3.1 High-level View

Bezoar is an automated VM-based recovery approach composed of the following modules (Figure 3.2): detection, memory monitor, log/replay/checkpoint and recovery. During normal execution the system logs all architectural non-deterministic events: interrupts and input events (Chapter 2, section 2.3). The memory monitor module tracks down how data from the network propagates into the memory (including registers and memory from external devices). Each memory byte will be associated with zero, or one, or more network sources. When an attack is detected, we discover its source and go back in time to the most recent checkpoint for recovery through replay.

The complete system execution is replayed until the first malicious input enters the system in a network packet. The malicious network input event is ignored (as all subsequent network packets coming from the same source) and the execution enters a semi-replay phase

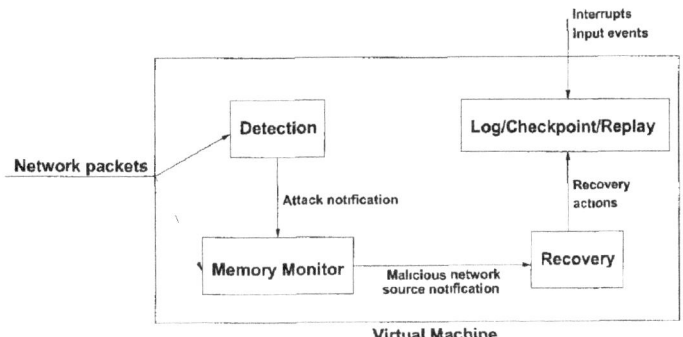

Figure 3.2. Bezoar modules.

In this phase the system executes partially in normal mode (without logging) and replay mode: it accepts and processes new events, but also replays non-malicious network packets so that their effects can be propagated up to the point the attack was originally detected. When the last non-malicious network packet is replayed, the system returns to its normal mode. During recovery (replay and semi-replay) the system does not send any data to the network.

3.3.2 Attack Detection

For attack detection we have employed the Biba's integrity model [59]. Integrity policies focus on integrity and accuracy of data rather than confidentiality [5]. A system is composed of subjects (active entities, such as processes and instructions), objects (passive entities, such as memory areas and files) and integrity levels. The higher the integrity level, the more reliable is the subject or the object. We have adopted Biba's low-water-mark policy, which has three rules [5]: (i) a subject cannot write into objects of higher integrity (no *write-up*), (ii) a subject's integrity level drops whenever it reads an object of lower integrity, and (iii) a subject cannot execute subjects of higher integrity (no *execute-up*).

We have used Minos [24] to detect zero-day control-flow hijacking attacks. It is a microarchitecture in which every 32-bit word of memory and general-purpose

```
high = 0;
for (loop = 0;loop<low; loop++)
{
    high++;
}
```

Figure 3.3. Defeating Minos information flow tracking.

register are augmented with an integrity bit, which is set by the kernel when it writes data into them. This bit is set to low or high, depending on the trust the kernel has for it. Data coming from the network are usually regarded as low integrity. Any control transfer involving untrusted data is considered a vulnerability, and a hardware exception traps to the VM whenever this occurs. Minos was prototyped in Bochs and can be implemented with negligible overheads [60].

It is important to point out that the information flow tracking of the Minos microarchitecture still can allow an attacker to remove an integrity label. The threat model we are able to deal with depends on the underlying information flow tracking for integrity. For honeypot applications or analysis in a laboratory environment, Minos is sufficient. For a production environment, where the attacker has knowledge about the information flow tracking, a system that makes stronger guarantees than Minos could be used, since we can employ any information flow tracking system in a modular way. As an example, the pseudo-code shown in Figure 3.3 illustrates how an attacker, that knows how Minos works, can insert the value of a low integrity variable into a high integrity one, without causing the high integrity variable to drop its integrity level. Minos does not check this indirect flow and, to the CPU, there was no arithmetic or data transfer instruction involving variables *high* and *low* while executing the loop. Thus, in Minos, there was no need to drop the integrity level of variable *high*. After the loop, *high* contains the same value as in *low*, which could be the address of a malicious instruction.

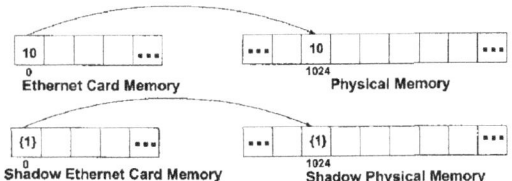

Figure 3.1. Memory Monitor - Example 1.

3.3.3 Memory Monitor

We use a memory monitor module running in the VM to keep track of how network bytes and their sources propagate into the system memory, including the register bank and memory from external devices, such as the network card. We make use of a shadow memory space and a shadow register bank to store information about the propagation of network bytes in our system. Each component of this shadow storage area has a 1:1 correspondence with the real component in the system architecture. The network byte propagation information is a set of integers representing the unique identification of their sources. Each network source is characterized by an IP address and a port number. In our approach, every time the VM Ethernet card receives a frame, we insert a new entry for the associated network source in our system if this is the first frame being received from the source. Then the frame bytes are stored into the network device memory and the identification number of their sources is stored into the shadow memory of the card at the same locations where the bytes were stored. The memory monitor can thus keep track of how these bytes propagate into the real system memory and registers as the CPU reads, moves and processes them. Wherever these bytes (and also their derived bytes) are stored, we can find their sources at our shadow memory space.

For example, let us suppose that byte 10 coming from network source 1 is copied from the Ethernet card to user space at memory location 1024 (Figure 3.4). Then the system receives byte 20 from network source 2 which gets stored into

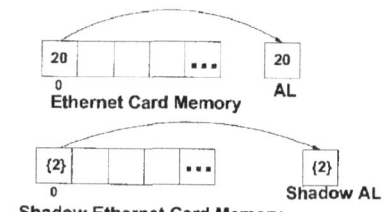
Figure 3.5. Memory Monitor - Example 2.

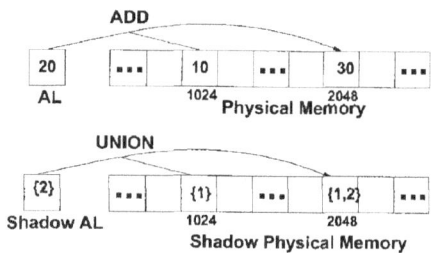
Figure 3.6. Memory Monitor - Example 3.

register AL (Figure 3.5). If the CPU executes an instruction that adds register AL to memory location 1024, and stores the result into memory location 2048, the final state of real and shadow memory will be as illustrated in Figure 3.6. Byte 30 at memory location 2048 is considered to be originated from network sources 1 and 2, and we store the union of the network source set associated with bytes 10 and 20 to the shadow memory at location 2048. When a memory location is overwritten with a byte that is not derived from any network source, we simply remove the set of network source ID numbers that might be stored into the corresponding location in the shadow memory.

For each network source active in the system we keep a counter for the number of shadow memory locations storing its ID number. When this counter reaches zero, we remove the network source entry from our system. The lifespan of most network sources is very short [24], for example, data from a TCP connection will soon die in the system after the connection was severed and data from new connections arrive.

3.3.4 Execution Checkpoint, Log and Replay

A post-attack recovery system needs to revert the attack actions in the victim host. We achieve this through replay, which allows us to go back in time and undo the effects of the attack in the system. We modified our previous log and replay framework ExecRecorder (Chapter 2) [8] with recovery capabilities: removing the execution of malicious events and performing the semi-replay phase.

3.3.5 Design and Implementation

We have used a Minos-enabled version of Bochs [27] as our VM and our proof-of-concept prototype was implemented as an extension to it. The architecture is the same as the one described in Chapter 2, Figure 2.2.

3.3.6 Identifying Network Inputs and Sources

Bezoar associates a network input event with its source as follows. When an Ethernet frame is read from the Ethernet device, we extract information about the source of this packet at the Network and Transport protocol levels. Bezoar maintains this information (source IP address and port number) in an object that we call **network source**. Each network source is uniquely identified by an ID (**network source ID**).

When Bezoar extracts the network source from an Ethernet frame it keeps its ID as the current network source ID in the system until a new packet from another network source arrives. After finishing reading the frame, the Ethernet device will interrupt the CPU, signaling that there are bytes to be read through port I/O. This interrupt may not be handled immediately by the CPU because it may have to handle higher-priority interrupts. When this interrupt is actually handled, the corresponding network bytes are read through port I/O. Thus, all network input events occurring until the next Ethernet frame arrives have as its network source ID the one kept as current by Bezoar. Figure 3.7 shows an example of this sequence

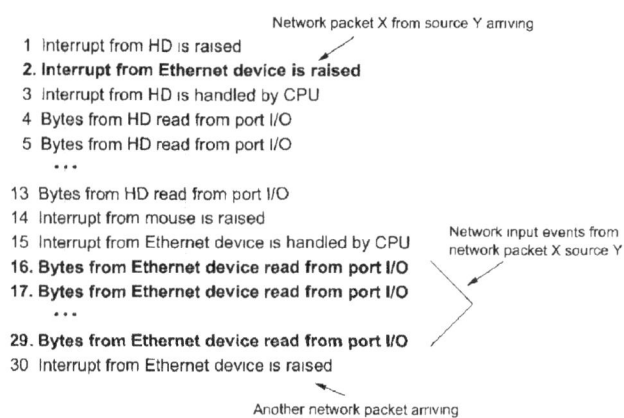

Figure 3.7. Identifying the source of network input events.

of events.

3.3.7 Identifying Malicious Network Inputs

If an attack is detected during the run, Bezoar discovers its malicious network source and initiates recovery by replaying the run without processing any event originated from the malicious source. During recovery the system does not send any data to the network. More specifically, at the detection point, Bezoar will inspect the shadow storage area (shadow memory and registers) the set of network sources stored at the same memory address or register from which the bogus value for the register EIP came from. This memory address or register was corrupted at some point by the exploit, and the value at this location is derived from the malicious bytes the attacker sent through a network packet (Figure 3.8). In all our experiments with real worms, the set of malicious network source directly associated with the attack was unitary.

3.3.8 Replaying to Recover from Attacks

Bezoar starts replaying the run knowing the ID of the malicious network source. During this replay for each network input event processed, Bezoar compares the

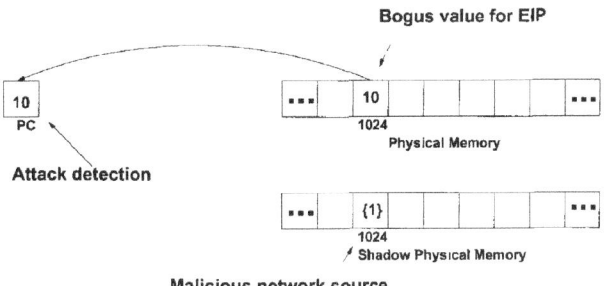

Figure 3.8. Discovering the malicious source.

network source ID for the event with the malicious network source ID. If they are different the event is replayed because it is non-malicious. If they are the same, this means that we are seeing the first malicious input event entering the system. The malicious event is not processed and the execution enters a semi-replay phase.

This phase represents our solution to the time-travel paradox discussed by Brown and Patterson [61] where this problem is compared to the metaphor of time-traveling portrayed in a famous movie: the protagonist goes back in time to repair some action that caused bad consequences to the present. After making the necessary changes, he inevitably affects the course of other actions and when he returns to the present, he sees a "modified" present that is consequence of the action he repaired along with the inevitably side-effects on other aspects of the past.

In our particular instance of time-travel, the repair action is to not process a malicious input as if it had never happened. We consider as malicious the stream of network packets coming from the discovered malicious network source. Up to the point that we see the first malicious input entering the system we can replay the run without worrying about side-effects. When we see the first malicious input entering the system we cannot just ignore all malicious inputs while replaying other events because certain non-malicious events may be dependent of some malicious

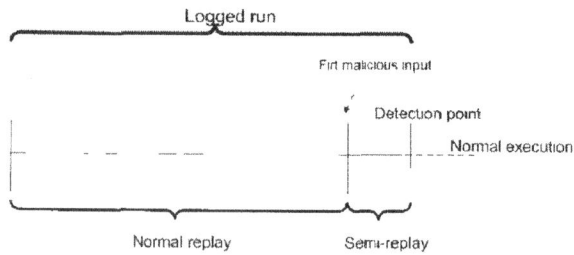

Figure 3.9. The semi-replay phase.

events and we could reach a situation of error or inconsistency. In our solution, when Bezoar sees the first malicious network input entering the system, we enter a semi-replay phase where the execution is partially in replay mode and in normal mode. Interrupts and input events not prescribed in the log file are processed, but network packets coming from the malicious source are rejected (these packets are not allowed to enter the system). Bezoar keeps replaying non-malicious network packets from the log file until all of these events are consumed. Our goal is to replay the system as if the malicious inputs had never occurred and bring the system state (memory, file system and processes) to a legitimate one: a state it would have reached had the attack never happened. When all non-malicious network packets in the log file are consumed, the system enters normal mode but continues to refuse any network packet coming from the malicious source. Figure 3.9 illustrates the semi-replay phase. It is important to point out that the system entropy after recovery will be different from what it was at the time the exploit was stopped. This is because during recovery, we remove from the system all malicious events, thus affecting the values at the system entropy pool.

3.4 Experimental Evaluation

Our experiments aimed at verifying if Bezoar could effectively recover a system after a control-flow hijacking attack and what performance overhead it incurred to

Exploit	OS	Port(s)	Class	bid[62]	Vuln. Discovery
DCOM RPC (Blaster)	Windows	135 TCP	Buffer Overflow	8205	Last Stage of Delirium
SQL Name Resolution (Slammer)	Windows	1434 UDP	Buffer Overflow	5311	David Litchfield
IIS (Code Red II)	Windows	80 TCP	Buffer Overflow	2880	eEye
wu-ftpd Format String	Linux	21 TCP	Format String	1387	tf8
rpc.statd (Ramen)	Linux	111 & 918 TCP	Format String	1480	Daniel Jacobwitz
innd	Linux	119 TCP	Buffer Overflow	1316	Michael Zalewski

Table 3.1. Worm exploits.

the VM. More specifically, we analyzed if Bezoar could (i) discover the source of the attack and distinguish malicious and non-malicious events and (ii) recover the server execution along with its complete state (file system, resources opened and interprocess communication).

We have selected six notorious exploits to test our approach. These exploits represent buffer overflow or format string vulnerabilities from Linux and Windows (Table 3.1). The Blaster worm exploits a buffer overflow in a RPC interface (DCOM vulnerability). The Slammer worm targets SQL server computers and exploits a stack buffer overflow vulnerability. Code Red II attacks the Microsoft IIS Server and exploits a buffer overflow vulnerability when a string in an HTTP GET request has its ASCII characters converted to UNICODE. The worm used for the *wu-ftpd* vulnerability was presented by Crandall and Chong [24], and exploits an error in the file globbing functionality. The *rpc.statd* worm attacks the NFS locking mechanism, and the *innd* attack targets the innd news server.

First we have tested if a Bezoar-enabled server could continue its execution after being attacked by any one of these exploits. For each exploit we have analyzed the server in three situations: idle, executing intensively its CPU, and running a Web server in a noisy campus network. For the second situation, we have selected for Linux the UnixBench [30] benchmark and for Windows an implementation of the Sieve of Eratosthenes. For the third situation we have used the Webstone 2.5

Exploit	Idle	CPU	Web server
DCOM RPC (Blaster)	0.368	0.401	0.415
SQL Name Resolution (Slammer)	0.096	0.406	0.594
IIS (Code Red II)	18.48	20.09	21.95
wu-ftpd Format String	35.68	37.38	17630
rpc.statd (Ramen)	0.769	0.888	0.872
innd	70.3	1172	9860

Table 3.2. Instructions between attack injection and detection (in millions).

benchmark [63] with 10 simultaneous clients fetching files of different sizes (from 500 bytes to 5 MB). The Web server running in our Linux guest OS is Apache 1.3 and in Windows is Apache 2.0. All the experiments were executed on a Pentium 4 SMP with 2 3.2 GHz CPU's and 1 GB of RAM. Also, each experiment was executed three times and its results averaged.

3.4.1 Effectiveness of the Recovery

We have attacked our Bezoar-enabled server with all the exploits presented in Table 3.1 and, for all of them, the server continued its execution without losing its full-system state. The microbenchmarks finalized successfully without any side-effects. In our experiments, as soon as the microbenchmark started executing, we launched the attack. Table 3.2 shows, for each exploit and three different types of workload, the number of instructions (in millions) executed from the time the malicious input entered the system through the network (attack injection) up to the time the control-flow hijacking attempt was detected.

The number of instructions executed between attack injection and attack detection determines the length of the semi-replay phase. The smaller the semi-replay phase, the less our recovery approach can cause side-effects or affect a client communicating with the server. As we have mentioned before, during semi-replay the system entropy is different from what it was during normal mode and under attack.

This is because during semi-replay we remove all malicious events from the execution, thus changing the values in the system entropy pool. If during semi-replay the communication between client and server depends on newly generated random numbers, for instance, an authentication key or a TCP initial sequence number, the numbers chosen during semi-replay will be different from those chosen when the system was under attack. In this situation the remote client will keep trying to communicate with the server with the old values (i.e., the values the server chose during normal mode and under attack) and this can cause a TCP connection to be reset or an authentication error.

In most of the attacks we analyzed, the size of the execution window between attack injection and attack detection is in the order of millions of instructions. In our VM, this lasts a few seconds, but for a CPU with a GHz clock speed this takes a small fraction of a second. The exceptions were the *innd* and *wu-ftpd* exploits whose semi-replay phase lasted in the order of billions of instructions when the host executed a Web server workload. This is due to the complexity of these exploits. The *innd* exploit has two steps: posting a message in a news group and then canceling it. The *wu-ftpd* exploit used in our tests is complex because it was specially developed to show the insecurity of detection techniques such as non-executable pages, return pointer protection and others [21]. It is important to point out that for the case where the system was running a Web server, our experiments were done in a local area network. In real-world attacks, that might take place across continents, it could be the case that the time between attack injection and detection lasts longer in clock time because of TCP round trip time. However, for this case, we could compress the idle periods in replay by identifying the idle loops and fast-forwarding the VMs CPU tick whenever we detected that the idle loop was being executed.

We have also noticed that the semi-replay phase executes fewer instructions

Exploit	Instructions Removed from Semi-Replay
DCOM RPC (Blaster)	6627
SQL Name Resolution (Slammer)	65
IIS (Code Red II)	130
wu-ftpd Format String	25758
rpc.statd (Ramen)	611
innd	10145

Table 3.3. Number of instructions removed from semi-replay.

than its corresponding window during normal execution (attack injection to attack detection). This is because during semi-replay we remove malicious events from the system and, consequently, all its corresponding instructions. Also, we enter normal mode as soon as we replay the last non-malicious network packet in the log file. In pure replay there may be other events to be processed after the last network packet is consumed. Table 3.3 shows, for each exploit, the number of instructions removed from the semi-replay phase.

We have also analyzed how Bezoar affected the communication of a client and a recovery-enabled server. We have considered the Web application where the Web server suffered a remote attack while servicing requests from remote clients. We have simulated 10 simultaneous clients fetching several files of different sizes (from 500 bytes to 5 MB). Each client kept making requests to the server for one minute. In each experiment, for each one of the exploits, as soon as the clients started making their requests, we launched the attack. Table 3.4 show the results we have obtained (average in three experiments per case).

We have analyzed all TCP packets exchanged by clients and server to find the cause for the errors in certain connections. We have observed that there was an error every time a new connection was started during the semi-replay phase. For example, suppose the server sends a TCP segment of type SYNACK to the client during the semi-replay phase. A SYNACK segment is part of the TCP three-way handshake and carries the initial server sequence number. During semi-replay,

Exploit	Number of Connections Tried	Number of Errors
DCOM RPC (Blaster)	7	1
SQL Name Resolution (Slammer)	7	2
IIS (Code Red II)	6	0
wu-ftpd Format String	34.75	3
rpc.statd (Ramen)	49	0
innd	70	2.25

Table 3.4. Number of client connections x Number of errors during recovery - Average.

the number chosen by the server is different from the number it chose during normal execution because the entropy of the system changed during semi-replay. As Bezoar replays all non-malicious network packets, the server, after sending the SYNACK, will receive an ACK segment from the client where its ACK number reflects the initial sequence number the server chose during normal execution. The recovered server, on the other hand, sees this mismatched ACK number as an error and resets the connection.

3.4.2 Performance

Figure 3.10 shows the performance overhead incurred by each one of Bezoar's modules and also all modules combined for five SPEC INT 2000 benchmarks. The execution times were normalized to the execution time of the VM without any of Bezoar's modules.

The log and recovery modules incur negligible overhead (less than 1%), while the memory monitor component, which tracks down the source of network bytes in the system, incurs an average slowdown of 1.4X. The total recovery time will be slightly less than the length of the checkpointed window because the semi-replay phase executes less instructions than its corresponding phase in normal mode.

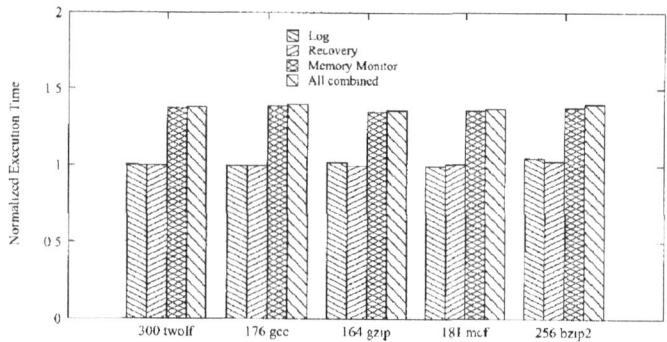

Figure 3.10. Performance overhead - SPEC INT benchmarks.

3.5 Discussion

There are some situations in which our approach cannot guarantee recovery. Currently our approach always goes back to the most recent checkpoint and, depending on the length of the run, the malicious input may have entered the system earlier. Also, Bezoar cannot guarantee recovery if the attack involves different TCP connections, or if the memory monitor component finds more than one network source associated with the attack. Another limitation is that we currently do not recover TCP connections that started during semi-replay and failed because the entropy of the system changed after recovery. We leave all these issues to future work.

Although we have described Bezoar by employing the tracking of network bytes during normal mode (with 1.4X slowdown), the system can also achieve recovery by enabling the memory monitor component only after an attack has been detected. This would require two replays of the checkpointed run. The first one would run tracking the source of network bytes to discover the malicious source, and the second replay would drop the malicious packets and perform the semi-replay phase. This configuration doubles the recovery time. Depending on the length of the checkpointed run it may be advantageous to avoid the 1.4X slowdown in execution time during normal mode.

3.6 Conclusions

In this chapter we presented Bezoar, an automated full-system VM-based approach to recover from zero-day control-flow hijacking attacks. It monitors the propagation of network bytes in the system and, after a control-flow hijacking attack has been detected, replays the checkpointed run while ignoring inputs from the malicious source.

Our approach is promising because for all exploits analyzed we could discover the source of the attack. The execution and the complete system state of the Bezoar-enabled server were recovered for all cases, and the communication between the server and any non-malicious client during the attack was recovered for most TCP connections. Only the connections that depended on the entropy of the system being the same as it was at the time the attack was stopped could not be recovered. This was due to the challenge of trying to control entropy (an OS concept) from a security solution operating at the architecture level. This prompted us to believe that if we could add active collaboration between the guest OS and the VM, we could bridge the semantic gap between these layers. In the next chapter we discuss in detail how we addressed it.

Bezoar incurs little overhead to the VM: less than 1% for the log and recovery components and an average of 1.1X slowdown for the memory monitor component that tracks down network bytes in the system. We expect that implementing Bezoar in a low-overhead VM will make it practical for use in real, online systems. Future work related to our post-attack recovery strategy is described in Chapter 6.

Chapter 4

Protecting Kernel Code and Data with a Virtualization-Aware Collaborative Operating System

4.1 About this Chapter

In this chapter we introduce collaboration between the guest OS and the VM as an efficient way to bridge the semantic gap between these layers [64]. We challenge the traditional VM usage model that advocates placing security mechanisms in a trusted VM layer and letting the untrusted guest OS run unaware of the presence of virtualization. We employ collaboration in a case study to protect the integrity of an OS kernel code and data. We have implemented a proof-of-concept prototype in Linux and have successfully tested it against 6 rootkits (including a non-control data attack) and 4 real-world benign LKM/drivers. All rootkits were prevented from corrupting kernel space and no false positive was triggered for benign modules. Performance measurements show that the overhead to the VM for the OS-VM communication is low (7% on average). The main overhead is due to the memory monitoring module inside the VM: 1.38X alone and 1.46X when combined with the OS-VM communication.

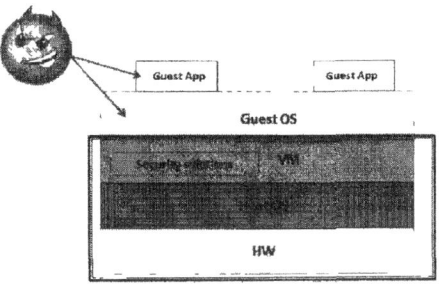

Figure 4.1. Traditional VM usage model.

4.2 Introduction

Although VMs have been used for timesharing capabilities (and even security) since the 1970s, during the last decade their topic has seen renewed interest for security-related applications such as intrusion detection systems, malware analyzers, secure logging and post-attack recovery using replay capabilities. This popularity can be attributed to the various desirable properties VMs have to offer [65]: environment isolation, OS and architecture extensibility, hardware multiplexing and easy manipulation of its internal state.

The traditional usage model, proposed by Chen and Noble in a seminal paper in 2001 [66], advocates placing security mechanisms in the VM layer (which has complete control of the system resources) and letting the guest OS run unaware of the presence of virtualization. The threat model assumed is that the VM is trustworthy and the OS running on top of it can be easily compromised by malware (Figure 4.1). The vast majority of security solutions presented in the literature employing virtualization [16, 21, 32, 67 76] does not count on the guest OS to enhance system security. A few solutions addressing active monitoring [77 79] employ limited OS involvement, which is accomplished by the introduction of hooks in certain points of kernel code. It does not represent, however, an active and on-the-fly OS collaboration.

In this work we challenge the use of this traditional model for security applications that rely heavily on inspecting or manipulating OS abstractions. First, security services operating only at one specific level of abstraction (*e.g.*, the architecture level represented by a VM) are limited due to the semantic gap between the VM layer and the requirements of the security solution to be provided, which usually involve manipulation and knowledge of information at several levels of abstraction to be effective. For example, a security solution operating only at the VM level and aiming to preserve the integrity of an OS code and data against attacks can access only architecture-level objects, such as CPU, physical memory, registers and external devices. However, in order to protect the OS kernel, such solution will need to manipulate OS-level objects such as processes, files and allocated memory areas (Figure 1.2).

Further, building a fully transparent, secure and isolated VM (for defense solutions and also for malicious actions) may be fundamentally infeasible, as shown by recent work [80]. Malware can present itself as a VM rootkit to conceal its actions and effects [81, 82], and detect it is running in a VM and change its behavior accordingly [83]. VMs can be detected because the majority of them have not been designed with transparency as a requirement [83], and because they interface with the Intel IA-32 ISA, which is not fully virtualizable [81]. Even with the proposed architecture extensions to make x86 processors fully virtualizable, virtual and physical environments differ in their timing characteristics and hardware configurations. For example, virtual environments will present large variances in the execution time of certain instructions when compared to native hardware. In the long run, attackers will operate regardless of the presence of VMs [80].

We believe that for certain types of security requirements a virtualization-aware and collaborative guest OS can provide, together with the VM layer, fine-grained, flexible and stronger system protection. The advantage of this model is combining

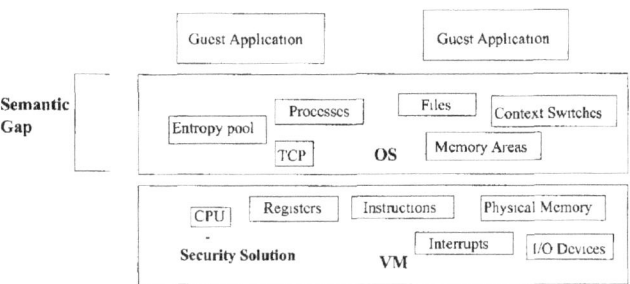

Figure 1.2. Semantic gap.

the isolation and hardware extensibility properties of VMs with the best possible view of OS objects in an architecture where the guest OS and the VM actively interface to achieve the requirements of a security solution. The first requirement of this architecture and interface is that an attacker should not compromise the integrity (code and data structures) of the collaborating OS running on top of the VM layer. Also, an attacker should not be able to modify the information exchanged between the guest OS and VM while interfacing with one another.

This chapter proposes a collaboration model between a virtualization-aware guest operating system and a VM layer to protect the OS kernel code and data integrity. In our solution we employ a relaxed version of Biba's integrity model [59] (Low-water-mark policy) on an architecture where a virtualization-aware guest OS and a VM layer can actively collaborate to enforce the model. All attempted writes into kernel code and data segments are checked for validity at the architectural level. If the instruction attempting the write operation is low integrity and the memory locations in kernel code or data segments to be written are of higher integrity, the write is aborted with the VM issuing a General Protection fault. This terminates the process associated with the write operation, but allows the system to continue its execution with its integrity preserved.

In our integrity model we enforce that no subject can write objects of higher integrity (first rule of Biba's low-water-mark policy). We adopt two levels of integrity: trusted (high) and untrusted (low). Immediately after boot, we consider everything in kernel and user space as high integrity (establishment time). After the establishment time, we consider every byte arriving in the system through the network (main vector for attacks) as low integrity, and we have the guest OS and VM to keep track of how these bytes propagate into the system at the architectural (memory and registers) and OS levels (files, processes, modules and dynamically allocated areas). This dual-layer of memory monitoring bridges the semantic gap between tracing low integrity objects at the OS-level and architecture-level.

This chapter presents the following contributions:

- We present a virtualization-aware architecture and interface where the OS and a VM layer collaborate to prevent integrity violations in kernel code and data by enforcing the first rule of Biba's low-water-mark policy: no *write-up*, i.e., a low integrity subject does not write into a higher integrity object).

- We implement this approach as a proof-of-concept prototype using the Bochs Intel IA-32 emulator and Linux;

- We evaluate this prototype against 6 rootkits and 1 benign loadable kernel modules

The rest of the chapter is organized as follows. Section 4.3 discusses attacks on kernel integrity. Section 4.4 addresses our integrity model and section 4.5 describes the high-level view of our approach. Section 4.6 details the design and implementation of our architecture. Section 4.7 presents our experimental evaluation. In section 4.8 we discuss the limitations of this work. Our conclusions are presented in section 4.9.

4.3 Attacks on Kernel Integrity

Rootkits are a type of stealthy malware that enable attackers who have gained administrator privileges or access to the system (usually through vulnerability exploitation or social engineering) to hide or covertly perform malicious actions and maintain control of the system. They have been increasingly used by attackers bundled with other type of malware such as Trojans, droppers, bots and key loggers. Kernel rootkits can corrupt, change, influence and add malicious behavior to the OS kernel which compromises the integrity of the entire system.

Preventing, detecting and recovering from kernel rootkit attacks is difficult given the complexity of kernel code and the great number and variety of its data structures. This complexity makes it harder to determine known kernel good states usually employed in defense approaches. Attacks in the kernel can succeed not only by adding or changing kernel code or altering its control flow, but also by tampering with certain key non-control data structures. For instance, recent work [85] has shown that an attacker can degrade system performance just by tampering with the value of the *zone_table* data structure and can compromise kernel security functions by contaminating the entropy pool. Further, an attacker has several avenues to compromise kernel integrity: vulnerabilities in kernel code, abuse of interfaces such as */dev/kmem* [86], malicious loadable kernel modules (LKM), hardware [87], virtual machines [81, 82] and social engineering.

The solutions for kernel integrity defense presented in the literature focus mainly on detection [70, 75, 88 92] and most of them are not effective for detecting attacks that involve value manipulation of kernel data structures [85]. The approaches addressing prevention rely on code authentication [69, 71] or security policies made by an expert that may fail to consider all possible data structures that can be abused by an attacker [67, 73, 74].

4.4 The Integrity Model

Our defense approach against violations in kernel space has the following goals: (i) prevent malicious tampering against kernel code and data, (ii) for any given attempt to violate kernel integrity, identify the process, file or kernel module associated and also the source of the attack, (iii) after a violation attempt, terminate the associated process or module, while allowing the system to continue execution normally, (iv) introduce minimal changes to the OS code and to the VM architecture, while keeping the performance overhead as low as possible.

In our solution we have adopted a relaxed version of Biba's low-water-mark policy [5, 59] described in section 3.3.2 from Chapter 3. In our approach, the subjects are instructions at the architectural level and processes and functions of loadable kernel modules (LKMs) at the OS level. The objects are the kernel code and data segments at the architectural level and files and dynamically allocated kernel memory at the OS level (Figure 4.4). In this model, the first rule (no *write-up*) states that a subject at a given level of integrity must not write any objects of higher integrity level. Adopting the other rules (no *read-down* and no *execute-up*) in the context of OSes is not practical given their design, system calls, interrupt handlers and some kernel threads, which we can consider as high integrity subjects, need to access objects at various levels of integrity in the course of their execution. For example, it is perfectly acceptable that a low integrity subject (*e.g.*, a process) invokes a high integrity subject such an OS function like *malloc* to allocate memory (violation of the no *execute-up* rule). Further, a high integrity subject, such as kernel function, may read low integrity data (parameters) without dropping its integrity level. The no *write-up* rule, on the other hand, if properly enforced, can protect kernel code and data from corruption and integrity violation (writes by subjects of lower integrity).

Immediately after the boot sequence (establishment time) we consider every-

Figure 4.3. Integrity model: subjects and objects.

thing in system memory and file system as trusted or high integrity. After the establishment time, we consider as untrusted or low integrity every byte arriving in the system from the Ethernet device. We have adopted this definition of trust because the network is the main attack vector for several types of malware.

4.5 High-Level View

In our approach, to preserve kernel integrity, we need to keep track of low integrity data at the guest OS and the architectural level. To accomplish this we need close cooperation between the guest OS and VM: the VM cannot track down the propagation of low integrity information at the level of files, processes and modules, as they are subjects managed by the guest OS. The guest OS, on the other hand, cannot manage the propagation of untrusted bytes inside memory and registers. For the remainder of this chapter we are going to use the terms guest OS and OS interchangeably, unless otherwise specified in the text.

At the architectural level (VM) we keep track of how network bytes propagate into the system memory (including registers and memory from external devices). At the OS level we keep track of files that are written with low integrity bytes, processes and modules whose object code is low integrity and areas in the kernel allocated by low integrity subjects.

The OS down-calls the VM (to request or pass information to it) through an unused software interrupt (vector 15, reserved by Intel). A software interrupt is a programmed exception and, as the name suggests, occurs at the request of the programmer. They are triggered by the **INT** family of instructions from the Intel IA-32 ISA [93] and are handled as a trap. A trap is handled by the control unit and is reported immediately after the instruction that caused it and when the OS regains control, the program that issued the INT instruction continues normally [26, 91]. System calls, for instance, are implemented as traps.

In our architecture, this unused software interrupt is called by the OS at certain points during system execution to request information from the VM about the integrity level of its objects and to pass information to the VM regarding boundaries of data structures and implicit propagation of low integrity data among OS objects that should be mirrored at the architectural level. The request parameters are passed in general purpose registers in an approach similar to what is done for system calls. Upon executing such instruction, the CPU invokes the OS Request Manager module to treat the request. Depending on the request type, this module reads the parameters from general purpose registers and may need to inspect internal VM state to serve the OS request. It returns information to the OS by writing into general purpose registers. Figure 1.4 illustrates this communication. The VM notifies the OS about integrity violations using exceptions.

Some VM-based introspective systems [77, 78] employ a technique known as *active monitoring* to improve the security of a system. In such a system hooks are placed inside the monitored system and when the execution reaches the hook, it is interrupted and control is passed to the security module. Active monitoring helps bridging the semantic gap between a VM layer and a security application. However, it does not take advantage of an active OS collaboration. In our approach the OS actively requests information from the VM or instruct it regarding boundaries of data structures that need to be protected. Also, approaches where the underlying host OS updates the VM memory directly would not allow the VM to gain fine-grained information, in a timely fashion, about the guest OS data structures.

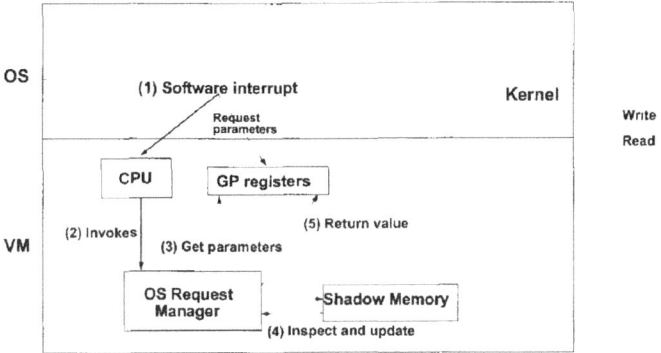
Figure 4.4. OS-VM communication.

4.5.1 Tracking Low Integrity at the VM Level

To keep track of how network bytes and their sources propagate into the system memory, we have used the same memory monitor module described in Chapter 3, section 3.3.3.

4.5.2 Tracking Low Integrity Data at the OS Level

4.5.2.1 Files

Every time a file is written, the OS down-calls the VM to request information about the integrity level of the bytes that are about to be written. Upon receiving the OS request, the VM checks in its shadow memory, at the location where these bytes are stored, whether or not that location is associated with a network source. A memory location that is associated with a network source is considered low-integrity. If at least one of the bytes is low integrity, the VM informs the OS that the file will be written with low integrity data.

4.5.2.2 Processes

In our architecture we set the integrity level of a process based on the integrity of its executable file. When the operating system is preparing to execute a certain process, it also checks the integrity level of the file carrying the process object code. If the file is low integrity, the process integrity level is also set to low.

4.5.2.3 Modules

The OS manages module integrity in a way similar to the method for processes: when a module is created, the OS checks the integrity level of the file storing the module's object code. If the file integrity is low, so will be the module's. In this case, the OS requests the VM to set the memory region holding the module's object code as low integrity in its shadow memory. This down-call is necessary because the bytes from the module binary file could have returned to the file system and upon re-entering main memory the VM will not have information about their integrity level.

4.5.2.4 Dynamically Allocated Kernel Memory

The OS also needs to inform the VM whenever a low integrity subject (a kernel thread or a module function) allocates kernel memory, so that the VM can set the corresponding memory range in its shadow memory as low integrity. As these areas should be considered low integrity, writes by low integrity objects into them should not be considered a kernel integrity violation. This allows a low integrity subject to be able to write into their own allocated kernel memory regions. Upon receiving the information about the new low integrity kernel memory region, the VM updates its shadow memory accordingly, *i.e.*, the allocated memory region is marked as low integrity, and is associated to the same network source belonging to the subject that allocated the area. When kernel memory is freed, the OS down-calls the VM, so that the integrity level of such areas are set to high.

4.6 Design and Implementation

We have implemented a proof-of-concept prototype for this architecture using the Bochs Intel IA-32 emulator [27] as our hosted VM and Linux 2.4.21 as the collaborating guest OS. We have used version 2.4.21 to maintain compatibility with previous work. The host OS is also Linux (2.6). Our prototype was implemented as extensions to Bochs and minimal changes to the collaborating OS. The hosted VM architecture is the same as the one described in Chapter 2, section 2.5.1

4.6.1 Writes into Kernel Space

Kernel space can be written by the following subjects: kernel threads and functions (including those from loadable kernel modules and system calls), processes in user space through the */dev/kmem* or */dev/mem* interface [1], and malware exploiting vulnerabilities in kernel code that allow direct injection of network bytes into kernel memory.

At the architectural level kernel space is composed of three distinct segments: kernel CS (code), kernel DS (data) and kernel SS (stack). In our defense approach we only enforce Biba's no *write-up* rule in kernel code and data segments. The stack segment is expected to be written by low integrity subjects executing at kernel level. For example, parameters of system calls may be low integrity but, as long as the system calls themselves are not corrupted, this does not threaten kernel integrity. System calls are intended to work on behalf of any given user process at any integrity level. Also, parameters of exported kernel functions may be low integrity if the module or kernel thread invoking the function is low integrity. Again, as long as the high integrity kernel function is not corrupted, the fact that it deals with low integrity parameters does not represent a breach in our integrity model because the data structures managed and changed inside these functions (and also inside system calls) are those expected to be changed in the course of system execution. For instance, when a low integrity kernel module registers itself as a character device in the system, it uses the exported kernel function *register_chrdev* (Linux 2.4.21). Inside this function, the kernel data structure *chrdevs* is updated with all

---

[1] These are character device files that allow privileged processes to access any physical page in the system by opening one of these device files and seeking a certain virtual (for */dev/kmem*) or physical address (*/dev/mem*) [86]. It is supposed to be used for debugging and quick kernel hacks but has been abused by attackers as a mechanism to inject malicious code or to tamper with kernel data structures.

the information about the new character device: major number, name and its file operations. In this case, the function will be invoked from a low integrity module, and consequently, the parameters passed will be low integrity as well.

On the other hand, data structures such as the system call table, processes list, zone table and entropy pool are not expected to be directly written. In fact, the kernel does not export functions that provide services for manipulating them. We are not arguing that it is not possible to conceive an attack using exported kernel functions. However, it would be much more difficult and would require much more ingenuity from the attacker.

4.6.2 File Integrity

In our approach we have considered that a file is low integrity if at least one of its bytes is untrusted (i.e., came from a network source). When a file is created into the system, the system call *sys_write* is invoked receiving as one of its parameters a buffer with the set of bytes to be written into the file.

To discover the integrity level of these bytes, the OS issues the software interrupt 15 passing as parameters the type of the OS-VM request (FILE_INTEGRITY) and the number of bytes to be written into the file. To answer such request, the VM needs to inspect in its shadow memory the integrity level of these bytes at the physical location where they are stored. If at least of one these bytes is associated with a network source in the shadow memory, the VM returns 1 to the OS via EAX register and 0, otherwise.

To inspect its shadow memory the VM needs to know the physical address of the buffer containing the bytes. As the OS deals with virtual addresses, our approach to obtain the corresponding physical addresses is as follows. After receiving the first software interrupt 15, the VM expects the OS to pass the initial address of the buffer by using a **DEC** instruction [93] with the first byte of the buffer as a parameter. Upon receiving such instruction immediately after the software interrupt 15 (FILE_INTEGRITY), the OS Request Manager interprets it not as a regular **DEC** instruction, but as a mechanism to obtain the address of the buffer to be written into a certain file. From the architecture point of view, this instruction behaves like a **NOP** [93]. While decoding the **DEC** instruction, the VM obtains the segment selector and the offset inside this segment where the first byte in the buffer is stored. From the segment and offset, the OS Request Manager computes the physical address and uses it to inspect its shadow memory to discover the file integrity. This solution is not ideal because the OS needs to down-call the VM via the software interrupt instruction **INT** [93] to inform the VM that the next **DEC** instruction should be interpreted as a mechanism to decode the virtual address (passed as a parameter) into a physical address. We actually needed a software interrupt instruction that supported parameters in memory. The Intel x86 **INT** family of instructions currently does not support this. A solution could be to extend the instruction set with a new instruction to perform this task. However, for compatibility reasons, we have chosen not to change the instruction set architecture.

After the **DEC** instruction, the OS issues another software interrupt 15 of type FILE_INTEGRITY exit mode and expects the VM to return in the EAX register 1 if the file is low integrity and 0 if it is high integrity. If the file is low integrity the OS marks its inode accordingly (new field *tainted* added to inode struct [91]). Also, in this case, the OS passes to the VM the file inode number and the file name so that the VM can maintain the information about low integrity files at the architectural level. This is useful for keeping track of kernel modules integrity. The OS passes this information to the VM in the same way that it does to communicate the size and the address of the buffer to be written into the file.

To keep track of the propagation of data from low integrity files even if their bytes leave the system memory (if the file is closed and later reopened), we introduce the following modifications into the *sys_read* system call. For each invocation of this system call, we check if the file from where bytes are going to be read to is low integrity. If the file is low integrity, the OS asks the VM to mark the region of memory where these bytes are going to be read as low integrity.

4.6.3 Process Integrity

At the OS level, we monitor process integrity as follows. Inside the *do_execve* function we check the integrity level of the process executable file by inspecting its inode's tainted field. If the file is low integrity we mark the new field *tainted* of the process's *task_struct* (Linux 2.1.21) as low integrity.

4.6.4 Module Integrity

A module is linked into the kernel by executing the user-space program *insmod* which performs the following operations: (i) reads from the command line the file name, (ii) locates the file that contains the module object code, (iii) computes the size of the area needed to store the module, (iv) invokes the *sys_create_module* system call to allocate kernel memory to store the module, and (v) invokes the *sys_init_module* system call to copy the relocated object code to kernel space and to call the module's initialization function.

As we can observe, the copy of bytes from the module object code file to kernel space will be done by code inside a system call that will obviously not make use of *sys_read*. Thus, to propagate their integrity level into the system memory, our approach leverages the fact that the VM maintains information about all low integrity files in the system (inode number and filename). The OS first asks the VM the integrity level of the module's object file. If it is low integrity the OS will ask the VM to mark, at the architectural level, the memory region storing the module object code as low integrity. This is necessary because the VM may not have information about the integrity level of the memory region holding the object code: the file bytes could, at a certain point, have left main memory and returned to the file system.

This OS-VM down-call proceeds inside system call *create_module*, immediately after memory is allocated to hold the module's object code. The OS issues a software interrupt 15 passing as parameters this type of request (MODULE_INTEGRITY) and the length of the module's name. Then, the OS issues a **DEC** instruction passing as a parameter the first byte of the buffer containing the module's name. The VM obtains the module's name and searches among the low integrity files the name of the module object file. If it finds the file, this means that it is low integrity. The OS obtains a final answer from the VM by issuing another software interrupt 15 passing as a parameter the type of the request (MODULE_INTEGRITY, exit). The VM returns in register EAX 1, if the module file is low integrity, and 0 otherwise.

If the module file is low integrity the OS needs to request the VM to mark as low integrity the memory region allocated for the module object code. The OS issues another software interrupt 15 passing as parameters the type of the request (MODULE_ADDRESS) and the size of the module. Then the OS issues a **DEC** instruction to pass the address of the first byte in the allocated area to the VM in the same way it did when it passed the module file name. The OS Request Manager inside the VM computes the physical address of the first byte in the allocated area and marks the whole memory region holding the module's object code as low integrity. The network source associated with this memory range is the same that is associated with the low integrity module file.

In our prototype we have used the name of the module file to identify a file at the VM level. The file inode number would have been a better choice because it uniquely identifies a file at the OS level. However, a module file inode number is not available to the module-related system calls (*sys_create_module* and *sys_init_module*). One way to address this limitation is to modify the system call *sys_create_module* so that it can receive the module file descriptor instead of the file name. This would require modifications to the *insmod* program so that it can pass the file descriptor to *sys_create_module* as a parameter. We leave this as future work.

4.6.5 Integrity of Dynamically Allocated Kernel Areas

Kernel areas dynamically allocated by low integrity modules or kernel threads should be considered low integrity. The kernel provides interfaces for memory allocation with or without page-size granularity [26]. The preferred choice for memory allocation inside the kernel is through the *kmalloc* interface, which requires the size in bytes of the required area passed as a parameter. The memory region allocated is physically continuous. The *vmalloc* interface is similar but the allocated area is only virtually contiguous. In this prototype we only keep track of the integrity level of areas allocated using *kmalloc* or *vmalloc* interfaces. However, extending this approach for page-size granularity interfaces such as *alloc_pages*, *__get_free_pages*, etc... is straightforward and we leave as future work.

Inside *kmalloc* and *vmalloc* (allocation interfaces), after the requested area is allocated, the OS obtains the address of the function that invoked it. Then it passes this address to the VM and asks it about the integrity level of the calling function. If the calling function is low integrity, the OS asks the VM to mark the allocated region as low integrity as well. In our implementation, we obtain the address of the function calling these allocation interfaces using the *__builtin_return_address* gcc hack to the linux kernel [95]. The OS issues a software interrupt 15 passing as a parameter the type of this request (ALLOC_AREA_CALLING_FUNC) and the size of the allocated area. Following, it issues a **DEC** instruction to pass to the VM the address of the calling function. The VM retrieves this address and checks in its shadow memory its integrity level. If it is low integrity, the allocated area should be considered untrusted as well. If the function is untrusted, the OS issues another software interrupt 15 passing as a parameter the type of the request (ALLOC_AREA_ADDR) and a **DEC** instruction with the first byte of the allocated area as a parameter. The VM computes the initial physical address of the allocated area and marks this entire memory region as low integrity. The associated network source is the same as the one found for the low integrity calling function.

When kernel memory is freed through *kfree* or *vfree*, the OS down-calls the VM passing as parameters the type of the OS-VM request (FREE_ALLOC_AREA) and the number of bytes that are being freed. Then, it issues a **DEC** instruction to pass the address of the freed area. The VM removes any association with network sources for this memory region, meaning that the integrity level of this region is set to high.

4.6.6 Kernel Threads

Kernel threads are privileged processes that run in kernel mode and in kernel space and do not interact with users [94]. They are usually created during system start-up, but after the establishment time, they can be created inside a module function.

All kernel threads created during system start-up (establishment time) are considered high integrity. However, we need an approach to monitor the integrity level of kernel threads created after the establishment time. As we have mentioned in section 4.6.4, the OS requests the VM to mark all areas holding a low integrity module object code as untrusted. Consequently, if a kernel thread is initiated inside a low integrity module function, all memory areas containing its instructions are considered as low integrity in the system.

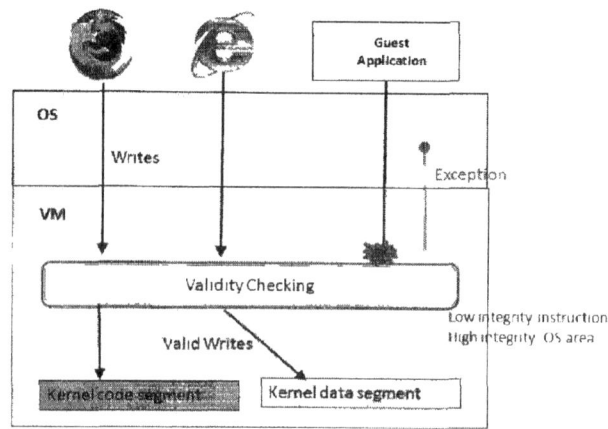

Figure 4.5. Enforcing the integrity model.

4.6.7 Enforcing the Integrity Model

We enforce our integrity model inside the VM as follows. At the architectural level the subjects can be represented as instructions executing in the CPU. The VM knows the integrity level of each instruction being executed at the CPU by inspecting it in its shadow memory at the location where the instruction is stored.

Every time the VM is executing an instruction that will cause a write into the kernel code or data segment it checks the instruction and memory areas integrity level. If the instruction is low integrity and the memory locations it will write into are high integrity areas of the kernel code or data segment, the VM aborts the write operation and issues a General Protection fault that terminates the offending process and allows the system to continue its execution normally with its integrity preserved. Figure 4.5 illustrates this sequence of events. The offending process and the network source associated with the aborted write are identified by the OS and the VM.

Rootkit/Attack	Attack vector	Parts of Kernel Affected	Non-control data
adore	LKM	system call table	No
rkit	LKM	system call table	No
knark	LKM	system call table, /proc interface and inet protocols	No
rial	LKM	system call table	No
SD & Devik	/dev/kmem injection	system call table	No
Resource wastage [85]	LKM	zone_table	Yes

Table 4.1. Kernel attacks.

Module/device	Description
ftpfs	Enhances Linux VFS with FTP volume mounting capabilities.
FISTGEN	Stackable file system templates and language.
kmw0.1.1	For modularizing Linux kernel functions even if they are not within a module.
frandom	Driver that implements a fast random number generator.

Table 4.2. Benign modules/devices.

4.7 Experimental Evaluation

In this section, we present the results of the experiments validating our architecture. We have implemented a proof-of-concept prototype using Linux 2.4.21 as our collaborating guest OS and Bochs IA-32 emulator as the VM layer. All experiments were executed on an Intel Core 2 Duo k6600 with 4GB of RAM. We had three metrics in this evaluation: (i) effectiveness against several types of kernel attacks, (ii) behavior against benign loadable kernel modules, and (iii) performance overhead.

4.7.1 Rootkit Attacks

We have selected 6 kernel attacks using two types of attack vectors (LKM and abuse of the *dev/kmem interface*) to test our approach (Table 4.1). To analyze the effectiveness of our solution against non-control data attacks [96], we have implemented a version of the Resource Wastage attack [85]. We have also selected 4 real-world benign kernel modules and devices to test our prototype against false positives (Table 4.2).

4.7.1.1 LKM Attacks

The most common attack vector for rootkits is to load themselves into the kernel through the LKM interface. A LKM is loaded into the kernel through the *insmod* program, which invokes system calls *sys_create_module* and *sys_init_module* for allocating kernel memory to hold the module and calling the module's init function. The module's init function is usually the place where rootkits presenting themselves as LKM tamper with kernel data structures, such as the system call table. The adore rootkit, for instance, replaces the entry of 15 system calls with its own malicious functions.

In our architecture when the first adore instruction attempts to write into the kernel memory holding the system call table, the VM detects that the instruction is low integrity, the memory region where it is trying to write is high integrity, and then issues a General Protection fault. The current process *insmod*, which is invoking adore init function, terminates and the system call table remains with its integrity preserved. The VM also reports the network source associated with the attack and the system execution can proceed normally afterwards. All rootkits that were tested and presented themselves as LKM were prevented from compromising the kernel the same way adore was

4.7.1.2 /dev/kmem Attacks

The /dev/kmem interface is a character device file that allows privileged processes in user mode to write directly into kernel memory as if it was a regular file [86]. The SD & Devik's rootkit [97] works by abusing this interface from user space. First it discovers the address of *kmalloc* and, through /dev/kmem, writes this address into an empty slot into the system call table. Then it calls *kmalloc* from user space as if it was a system call to allocate kernel memory. This memory is written with malicious code and the rootkit uses another empty slot from the system call table to store the address of this code (KINIT). When KINIT is invoked from user space it replaces other system call entries with pointers for malicious versions of them.

When the rootkit is executed in our architecture and calls KINIT, the VM detects that a low integrity instruction in kernel mode (instruction from KINIT in system call context) is attempting to write into a high integrity kernel memory region (system call table) and then issues a General Protection fault that terminates the rootkit process with message: *Got signal 11 while manipulating kernel!*.

Although we were able to prevent the rootkit from replacing system call entries with its malicious versions, kernel space was still corrupted because, during the first phase of the attack, the SD & Devik's rootkit was able to write into empty slots from the system call table by abusing /dev/kmem. This happened because our architecture currently only checks the validity of writes performed in kernel mode, when the segments loaded into DS, CS and SS registers are respectively kernel data, code and stack segments. The corruption of kernel space in the first phase of the attack (overwriting the two empty slots from the system call table with the addresses of *kmalloc* and KINIT) occurred in user mode when the data, code and stack segments being addressed by the CPU were those from a user process. Our solution to this issue was to check the integrity level of processes trying to open /dev/kmem for write operations. If the process integrity is low the open operation returns an error code. This prevents the rootkit (and also other rootkits that work by abusing this interface) from succeeding. It terminates with message *Can't open /dev/kmem for read/write*. The kernel integrity remains intact. Another solution would be to also check the validity of writes in user-mode for segments other than those loaded into the DS, CS and SS registers. We leave this alternative as future work.

4.7.1.3 Noncontrol Data Attack

The majority of rootkit attacks works by tampering with control data-structures, such as the system call table. The goal is to control user-level requests so as to hide or covertly perform malicious actions. Recent work [85], however, has shown a new class of kernel attacks targeting non-control kernel data structures. We have implemented a version of the Resource Wastage non-control data attack, presented by Arati et al. [85]. It causes resource wastage and degradation of system performance by generating artificial memory pressure. The attack tampers with zone marks in the normal memory zone. These zone marks are stored in the *zone_table* data structure. We have implemented a LKM that sets the *page_min* and *page_low* watermarks close to the value of *page_high*.

When loading this module in our architecture, the VM detects a low integrity instruction (from the module's init function) attempting to write into the high integrity data structure *zone_table*. As in the case of other LKM rootkits, the VM issues a General Protection fault that terminates *insmod* without loading the rootkit and maintaining the integrity of the *zone_table* data structure intact.

4.7.1.4 Benign LKM's

We have selected 4 benign modules/drivers (Table 4.2) from sourceforge.net [98] to test our architecture against false positives. During our tests, none of these modules violated our integrity model when loaded into the kernel.

It is important to point out that if a benign module tries to legitimately modify

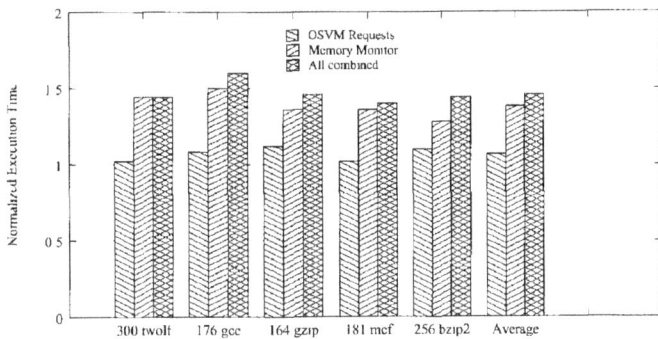

Figure 4.6. Performance overhead - SPEC CPU benchmarks.

areas of the kernel it does not own, for instance, adding a new system call, the VM will prevent such action and report it as a security violation the same way it does for malicious modules.

4.7.2 Performance Overhead

We have evaluated the performance overhead incurred by our architecture using a subset of the SPEC CINT2000 CPU benchmark suite [99] and also the system microbenchmarks from Unixbench [30]. The execution times were normalized to the execution time of the system without the OS Request Manager and memory monitor modules at the VM layer, and running on an unmodified Linux 2.4.21 guest OS.

Figure 4.6 shows the performance overhead incurred by the OS-VM communication, the memory monitor module inside the VM, and all modules combined for SPEC CINT2000 (CPU). The average overhead for the OS-VM communication in benchmarks exercising mainly the CPU is low, approximately 7%. The greatest overhead is caused by the shadow memory operations from the memory monitor module inside the VM, which caused a slowdown of 1.38X alone and 1.46X when combined with the OS-VM communication.

Figure 4.7 illustrates the overhead of the OS-VM communication for system mi-

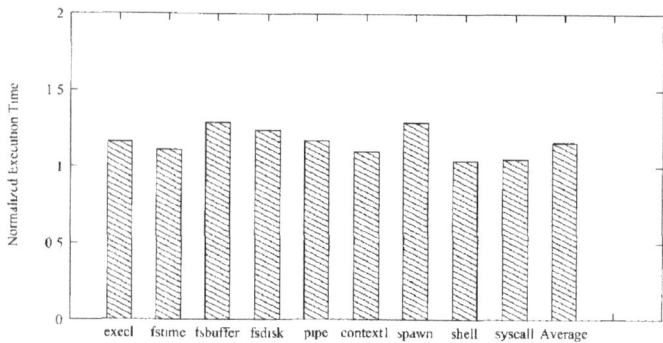

Figure 4.7. Performance overhead - Unixbench system microbenchmarks.

crobenchmarks (exercising OS functionalities such as file system, process creation and execution, pipe etc...) from Unixbench. On average, the system overhead is 16%, where the greatest performance penalties were observed for microbenchmarks exercising the file system and process creation. This is because their related system calls (sys_write, sys_read and sys_creeve) caused a OS-VM communication in most of their invocations.

4.8 Discussion

An attacker could try to compromise our system by issuing malicious OS-VM requests from inside a kernel module. For instance, she could request the VM to set as high integrity a memory area she controls through OS-VM request FREE_ALLOC_AREA. However, the VM can still control which subjects at OS level can perform down-calls by checking the integrity level of the instruction issuing the OS-VM request. Only high integrity instructions should be allowed to perform such requests.

Benign kernel modules coming from the network that modify kernel data structures without using an exported kernel function will have their actions reported as an integrity violation. This situation can be overcome if such modules are obtained from a network interface that the system considers high integrity or if they are installed in the system offline.

We currently do not provide protection against malicious DMA writes into kernel space or against any write operation originated from malicious peripheral devices. However, in many corporate settings one could expect several attacks arriving via infected USB keys. To address attacks arriving from a certain peripheral device we could consider low integrity all bytes coming from the device ports and employ the same approach we did for low integrity bytes coming from the network.

Also, we cannot offer protection against attacks that do not need to write into kernel space to succeed. One example are hardware supported rootkits, such as Cloaker [87]. Other type of attack that does not rely on writing into kernel space and that has been recently presented in the literature are soft timers attacks [100]. This type of attack works by installing a malicious soft timer into the system that can degrade its performance or steal sensitive information. Also, capturing VM-based rootkits [81, 82] is out of the scope of this work.

Finally we currently do not offer protection against low integrity subjects writing into low integrity objects that they do not own, for example, a low integrity module corrupting memory areas allocated by another low integrity module or kernel thread. We leave this issue as future work.

4.9 Conclusions

This chapter discussed how we challenged the traditional model of placing all security solutions inside a VM layer and having the guest OS run unaware of virtualization. We believe that for certain security requirements, an OS actively collaborating with a VM layer underneath is promising for fine-grained and stronger system protection.

We presented an approach employing virtualization-aware OS and VM collaboration to prevent violations against kernel code and data. Our proposed architecture was implemented as a proof-of-concept prototype and we adopted a relaxed version of Biba's integrity model, where we validated all attempted writes into kernel code and data segments. In our approach, if the instruction attempting the write operation was untrusted and the locations in kernel space to be written were high integrity, the VM aborted the write and issued a General Protection fault, terminating the offending process without any compromise to kernel integrity. The OS-VM direct interaction bridged the semantic gap between tracing low integrity objects at OS-level and architectural level.

We successfully tested our prototype against 6 rootkits (including a non-control data attack) and 4 real-world benign LKM/drivers. The average overhead to the VM for the OS-VM communication for CPU benchmarks was low, approximately 7%. The memory monitor module incurred a slowdown of 1.38X alone and 1.46X when combined with the OS-VM communication (SPEC CPU benchmarks). The overhead for the OS-VM communication for OS microbenchmarks was approximately 16% on average. We discuss future work in Chapter 6.

Chapter 5

Related Work

5.1 Virtual Machines and Security Solutions

First proposed by IBM in the 1960s for mainframe timesharing capabilities and used for security isolation and mediation in the 1970s, VMs have seen renewed interest since the beginning of the last decade in applications such as system isolation and migration, platform replication, concurrent execution of OSes, testing of new applications or OS features, among other uses [65].

In a seminal paper in 2001 [66], Chen and Noble proposed running applications and OSes inside a VM, where the services run without trusting or modifying the guest OS. They argued that VMs expose a narrow interface and can also be modified more easily when compared to OSes, and thus, are less susceptible to vulnerabilities. They suggested secure logging [16], intrusion detection or prevention [24, 32, 67, 71, 88], and environment migration as *killer* applications. Other suitable applications (not mentioned in the paper) include software authentication [72] and OS debugging [101]. The authors also pointed out the main disadvantage of this model: the semantic gap between the VM and the security services to be provided.

In [102], Garfinkel and Roseblum discussed security problems in VM environments. An example of challenge is that the rapid scaling of VM environments can make administration tasks complex in an organization and create a suitable environment for malware propagation. Roscoe et al. [103] raised the question of whether the combination of a modified VM and a legacy OS is better than simply focusing on building better OSes. They advocated the use of VMs only to support legacy applications, write drivers and redefine OS interfaces.

Although VMs have been marketed as a commonplace for detection and analysis of malware, malware can fight back by (i) presenting itself as a VM rootkit to conceal their actions and effects and (ii) detecting it is running inside a VM and changing its behavior accordingly [83]. Subvirt [81] is a VM-based rootkit installed underneath the OS. It boots and loads the existing OS in a VM and, while the user unknowingly works in that VM, a second, hidden VM performs malicious tasks. Blue Pill [82] traps a running instance of the OS into a VM, and acts as a hypervisor, with complete control of the computer.

VMs can be detected because the majority of them have not been designed with transparency as a requirement [83] and because they interface with the Intel IA-32 ISA, which is not fully virtualizable [81]. This architecture has seventeen instructions that are sensitive but unprivileged, i.e., their execution does not cause a trap to the VM but they can still reveal information about the architected state of the machine, for instance, the content of certain registers and the privilege level at which a program is running. This information can help a piece of code (malicious or not) to identify whether or not it is running inside a VM. For example, if a program which is supposed to run in kernel mode (ring 0) inspects the architected state of the machine using one of these sensitive but unprivileged instructions, and finds out that it is running actually in user mode (ring 3), it can infer that it is running inside a VM environment and not in native mode. Many argue that the

Intel [101] and AMD [105] architecture extensions to make x86 processors fully virtualizable will make it much more difficult to detect a VM. However, a recent paper [80] concludes that building a transparent VM is fundamentally infeasible. This is because VMs will tend to present slightly different hardware configurations and timing discrepancies when compared to physical environments. For example, virtual environments will present large variances in the execution time of certain instructions when compared to native hardware.

Lares [77] is an architecture for secure active monitoring which aims at bridging the semantic gap between VM and security applications. It consists of placing hooks inside the system being monitored and when the execution reaches the hook, it is interrupted and control is passed to the security application. The work shows design techniques that allow the installation of protected hooks into an untrusted VM. Panorama [78] is an example of a VM-based active monitoring system. It employs operating system-aware, whole system taint checking to analyze offline whether or not an unknown program is privacy-breaching, i.e., accesses data it is not supposed to. The analysis is done as follows. First sensitive information is introduced in the system in a way that is not intended to be used as input by the program under analysis. Then, Panorama checks, using taint tracking, whether this sensitive information has propagated into the program being analyzed and what it has done with it. Although active monitoring helps bridging the semantic gap between a VM layer and a security application, it does not take advantage of an active OS collaboration.

5.2 Kernel Integrity Defense

5.2.1 Prevention

The majority of works in the literature addressing prevention from kernel attacks focus on preventing unauthorized code from being executed and cannot guarantee protection against non-control data attacks [96], for instance, attacks that tamper with kernel data structures by directly injecting values into kernel memory through vulnerabilities, or by abusing of interface such as /dev/kmem.

Manitou [106] leverages a hypervisor to ensure that only authorized code run in the system. It uses per-page permission bits to ensure that any code contained in an executable page is authorized for execution. Code is authenticated by taking cryptographic hashes of the page content before execution.

Grizzard [107] proposes an approach to protect kernel integrity by monitoring its execution and validating its control flow. The kernel traps all dynamic branches into the VM before execution and they are verified for consistency with the kernel control flow.

SecVisor [69] is a hypervisor that protects OS code integrity by ensuring that only user-approved code executes in kernel mode. In this approach, users supply a policy that is checked against all code loaded into the kernel. It works by virtualizing physical memory, which allows that hardware protections be set over kernel memory. During system execution, the CPU refuses to execute any code that is not user-approved.

NICKLE [71] uses a memory shadowing scheme to prevent unauthorized code from executing in kernel mode. In this scheme a trusted VMM (virtual machine monitor) maintains a shadow copy of the main memory for a running VM and performs kernel code authentication so that only trusted code is copied into the shadow memory. During execution, instructions are fetched only from the shadow memory.

The solutions addressing prevention from non-control data attacks on kernel data structures rely on policies that explicitly consider the data structures to be protected. Given the great number and variety of them, these policies could be incomplete and fail to address all range of attacks.

Livewire [67] is a VM architecture whose policies protect certain parts of kernel space such as the code section and the system call table from being modified. Paladin [73] is a mechanism for detection and containment of rootkit attacks using VM technology. The approach relies on an administrator to specify access control policies for certain memory areas and system files. These policies protect files from being replaced and memory areas from being overwritten. A tracking mechanism creates a dependency tree based on the relationship between processes and the files they create. This tree is used by the containment algorithm to identify malicious processes and stop ongoing damage as soon as a violation is detected.

Xu et al. [74] propose a framework with an access control model for the specification of policies and an architecture to enforce kernel integrity. Kernel integrity policies are specified through an usage control model and the access to kernel code and data structures is controlled according to pre-defined access control policies.

Other prevention approaches include rootkit resistant disks, secure virtual architecture for memory, and code attestation. Butler et al. [108] introduce the idea of rootkit resistant disks (RRD) to prevent a compromised OS from infecting its disk image. At installation time, all binaries and configuration files are labeled as immutable. During normal system execution, the disk controller checks all write operations and those directed to immutable blocks are denied.

Secure Virtual Architecture (SVA) [109] is a virtual low-level typed instruction set that enforces a safe execution environment (memory safety, control-flow, type safety and sound analysis) for kernel code and its applications. It does not prevent, however, malicious code not exploring memory safety errors (buffer overflows, format strings, double frees) from corrupting the kernel and changing its behavior.

Code attestation techniques [72, 110-112] verify a piece of code before it gets loaded into the system. They do not protect kernel integrity against memory injection attacks.

5.2.2 Detection

The majority of works in the literature about kernel rootkits addresses detection. Copilot [89] is a kernel integrity monitor that uses a PCI add-in card to access memory instead of relying on the kernel to accomplish that. It uses MD5 hashes of kernel text, loadable kernel modules text and the contents of the system call table as its detection strategy: it periodically calculates these hashes during runtime and checks if results differ from known good states.

The same authors of Copilot [89] addressed its limitation of not being applicable to dynamic kernel data structures with an architecture that detects kernel violations by comparing the kernel state with a specification of a correct state done by an expert [90]. It includes a compiler that translates this specification into machine code that performs the checks. The fact that this architecture depends on a human to specify what a correct kernel state is may prevent it from capturing all types of kernel violations as the specification could be incomplete. The authors address this shortcoming with a technique that employ an approximation of control-flow graphs [113], called state-based control-flow integrity - SCFI), to periodically validate kernel state (each dynamically-computed branch is validated) during execution [75]. The main assumption is that rootkits add functionality to the system and if we know the paths (according to a control-flow graph) to be followed during software execution, we can detect when the kernel was violated. The implementation consists of a monitor kept separate from the kernel (as in Copilot [89] and the architecture presented in [90]). The state analysis is done in two steps. First the kernel code is checked. Then a garbage collection style traversal of the heap is done to find usable function pointers and verify they target valid code. This technique has the limitation of detecting only changes to kernel control flow. Attacks that tamper with kernel data structures (such as the Resource Wastage attack described by Arati et al. [85]) or that modify the kernel for short periods of time will succeed.

Strider GhostBuster [91] is a framework for ghostware (programs that hide their resources from the OS) based on a cross-view approach: hiding behavior is captured by comparing two snapshots of the same state at the same time but from two different points of view (one from the malware and the other not). GhostBuster first performs a high-level and a low-level scan of the resources in an inside-the-box solution. When the OS is infected with a ghostware, the high-level scan contains the untrusted view, the low-level scan contains the trusted view and their difference exposes the hidden resources. Lycosid [70] and VMWatcher [88] also use this cross-view principle to perform detection.

Gibraltar [92] is a tool to automatically detect rootkits modifying both control and non-control data. During a training phase the kernel execution is observed and invariants for kernel data structures are hypothesized. These invariants, which include properties of both control and non-control data, are used as specifications of the integrity of data structures during an enforcement phase. A violation in these invariants indicates the presence of rootkits.

Wang *et al.* [114] propose an approach that uses a rootkit detection program to discover hooks that could be used by kernel rootkits to avoid detection. The solution is based on instrumenting and recording control-flow transfer instructions.

5.3 Taint Tracking

Data flow tracking is referred to tagging data with some extra information that is propagated when this data is involved in some operation in the system. This extra information can be just a bit, for example, describing if the data is trusted or not, or can represent more complex information such as the source of the data. When the tagged information is just one bit it is called taint tracking. It has been used in security solutions to discover vulnerabilities and attacks when tainted data reaches a certain region of memory or is employed in a certain operation it was not supposed to [21, 37, 115-117]. Recent work have been focusing on improving the considerable runtime overhead of such approaches [38, 60, 118, 119].

TaintBochs [116] analyzes sensitive data lifetime in large software systems by tainting sensitive data with a 1-bit flag (*e.g.*, a password) whenever it enters the system and tracking it through a whole-system Pentium simulator. A shadow copy of memory and register is used to maintain tainted information. An analysis framework running offline uses information logged during system execution to trace tainted data to a program variable in a guest application and code propagating tainted data to a source file and line number.

Minos [24] is a microarchitecture to detect zero-day control-flow hijacking attacks that implements Biba's low-water-mark integrity policy [59] on individual words of data. Minos can detect attacks that corrupt control data to hijack program's control flow, where control data is considered any data that is loaded into the program counter (EIP) on a control flow transfer instruction or any data used to calculate it, for instance, return pointers, function pointers, jump targets, etc... In Minos, every 32-bit word of memory and general-purpose register is augmented with an integrity bit, which is set by the kernel when it writes data into them. This bit is set to low or high, depending on the trust the kernel has for it. Data coming from the network are usually regarded as low integrity. Any control transfer involving untrusted data is considered a vulnerability and a hardware exception traps to the kernel whenever this occurs.

Vogt *et al.* [117] present a solution to stop XSS attacks on the client side by tracking the flow of sensitive data inside the web browser. The goal is to enforce that a JavaScript program only send sensitive information to the site from which it is loaded. Their technique tracks sensitive data inside the browser's JavaScript engine. Whenever there is an attempt to leak information to a suspicious site, the user is informed and can prevent the transfer.

Panorama [78] employs operating system-aware, whole system taint checking to analyze offline whether or not an unknown program is privacy-breaching, *i.e.*, accesses data it is not supposed to do. SHIFT [118] is a low-overhead, software-based dynamic information-flow tracking that leverages the existing architectural support for speculative execution to track tainted data. The main idea is to treat tainted data as speculative state so taint tracking can be emulated using deferred exception tracking in microprocessors supporting speculative execution.

5.4 Deterministic Execution Replay

Instant Replay [20] is a deterministic replay for highly parallel programs to help the debugging process. During program execution the relative order of significant events is saved without recording any data associated with them. It is well-suited for replaying non-interactive applications.

Flashback [19] is a lightweight OS extension for software debugging that provides replay capabilities for an application. The main idea is to use shadow processes to capture the in-memory state of a process at a specific execution point and log the process interactions with the system. Our replay-based post attack analysis approach, ExecRecorder (Chapter 2), can replay the execution of an entire system in the same environment as it happened during log and does not require any changes in the OS. In Flashback, a replayed application may be executed in a different environment.

Rx [18] proposes an interesting technique to survive software failures by treating bugs as allergies: their manifestations can be avoided if the execution environment is changed. As in Flashback [19], it also uses shadow processes and the application replay happens in a changed environment.

FDR [23] is a low-overhead, full-system hardware recorder for software debugging. ExecRecorder is different from FDR in that it records architectural events inside a VM and does not address multiprocessors and DMA, while FDR is an actual design of a hardware recorder for multiprocessors. On the other hand, FDR only enables replay intervals of approximately one second, does not capture disk state, and does not provide a fully-implemented replayer. ExecRecorder allows replay windows of any length (provided that there is enough disk space to store the log files) and checkpoints the disk with copy-on-write.

BugNet [17] is a log and replay architecture for debugging. It records the register file contents at any point in time, and the load values that occur after that point. As it focuses on application level bugs, it cannot replay the full-system execution.

Dunlap *et al.* proposed ReVirt [16], a logging and replay system for analyzing intrusions that runs integrated with a VM and performs the logging in the host OS. After an attack, it can replay the whole VM process for analysis. ExecRecorder is different from ReVirt because our application requires fine-grained control and more flexibility in the choice of the guest OS. ExecRecorder does not require any changes in the guest or host OS. This is important because most Internet worms attack Windows, and many worms or attacks must be caught with a specific version of an OS. For our studies we needed to be able to run Windows XP, Windows 2000, Windows Whistler, as well as a variety of Linux, FreeBSD, and OpenBSD distributions, and interface with these using a Pentium hardware interface. ReVirt is implemented as a set of modifications in the host kernel and the guest OS must be ported to run on their VM (UMLinux). Second, ExecRecorder gives us complete control of a system under replay, which is necessary for addressing post-attack recovery and replay-based entropy control [34]. In post-attack recovery, where we want to recover the system by disabling particular effects of the attack and replaying the modified run, we needed full-control of each event being replayed. ReVirt, by just replaying the VM process from the perspective of the host OS, does not offer such level of control. While Bochs limits the performance of our implementation, we can do very sophisticated analysis without prohibitively affecting system performance by using the low-overhead ExecRecorder module, and then do the analysis during replay with much more expensive modules such as DACODA [25].

5.5 Recovery

Recovery has been addressed for a long time in areas such as databases and fault tolerance. The approaches used for databases aim to protect data integrity [120]. In the area of software fault tolerance the goals are to restart an application after a fault, or periodically rejuvenate a system to avoid a fault, or roll-back and/or replay an application after a fault. This can be achieved through reboot [121 123] and its variations [124, 125], software rejuvenation [126], and roll-back recovery and replay techniques [9, 10, 13, 127, 128]. These approaches cannot be directly applied for post-attack recovery because or they do not address availability, or may lose system state while recovering, or do not include a repair mechanism.

Just re-starting an application does not improve system availability and rolling-back the application to a previous checkpoint (without logging) and letting it continue will not allow nondeterministic events to be regenerated. Replaying the application from logged files without employing a repair mechanism will just cause the attack to be replayed again.

5.5.1 General Software Failures

In Rx [18] a program is rolled back and re-executed (possibly multiple times) in a modified environment when a bug is detected. Brown and Patterson [61] propose the 3R's (Rewind, Repair and Replay) model to address the problem of recovery from mistakes made by humans, where a human is also the entity responsible for detecting that an error has occurred and discovering how to repair it. BackDoor

[129] is a recovery system that detects when an OS is unresponsive and salvages critical software state in the OS memory so that another recovery node can take over the client sessions serviced by the failure node. Other related works are the runtime systems DieHard [130] that tolerates memory errors, and Exterminator [131] that detects and corrects heap-based memory errors.

5.5.2 Worms and Malware

Sweeper [50] extends Rx [18] with recovery capabilities and employs heavyweight post attack analysis using multiple replays to discover the source of an attack and to generate a filter for it. INDRA [51] is an architecture that asymmetrically configures the processor cores in a security hierarchy. High privileged cores isolated from the network monitor low privilege cores running network services. It does not recover the full-system state such as file system, interprocess communication and resources (files opened, processes spawned and pages allocated). Taser [132] recovers the file system after an attack using taint tracing. Sidiroglou et al. [52] use rescue points to restore the program execution after a fault, and STEM [53] provides recovery by having the faulty application return an error. DIRA [54] and failure oblivious computing [55] represent compiler solutions for the problem. Back to the Future [133] provides recovery from malware like Trojan horses, spyware and viruses. It detects an integrity violation when a trusted process is about to read untrusted data, and depends on a human to classify an application as trusted and untrusted.

5.6 Post-Attack Analysis

DACODA [25] is a tool that analyzes attacks using symbolic execution. It labels each byte coming from the network with a unique identifier and tracks these bytes in the system during their lifetime. When an attack is detected, DACODA provides information about it, such as processes involved, if the attack involved kernel or user processes, tokens that compose the attack trace and the predicates found.

Xu *et al.* [45] and COVERS [46] analyze the victim host memory and correlate attacks to network inputs to perform signature generation. Brumley *et al.* [47] analyze the semantic of a program to generate a filter for a certain vulnerability. Vigilante [18] is a containment approach where a set of hosts run instrumented software and, upon an attack, broadcast self-certifying alerts to other hosts.

Chapter 6

Conclusion

This dissertation presents virtual-machine based mechanisms and tools to prevent and respond to certain types of malware and attacks. It describes a replay-based approach to analyze attacks offline and to recover a host execution from control-flow hijacking attacks. The dissertation also challenges the traditional VM usage model where all security mechanisms reside in a trusted VM layer, which does not count on the guest OS (unaware of virtualization) to enhance system security. The approaches described in this dissertation could recover and analyze several attacks from notorious Internet worms and prevent tampering against operating system code and data from various types of kernel rootkits.

6.1 Summary

We began addressing response against cyber attacks in the context of virtual machine environments by proposing a replay-based approach for post-attack analysis and recovery. When we started, prevention and response against malware and attacks did not get as much attention as detection in security research. The majority of efforts in malware defense were in the area of detection (mainly detection and signature-based systems), meaning recognizing that an intrusion, infection or attack is taking place, and possibly stopping it afterwards. We proposed a full system replay for uniprocessors for offline post-attacks analysis where we could faithfully replay the execution of an entire system from a checkpoint with low performance/space overhead. Its main advantage compared to previous approaches was that it worked at the architectural level, which allowed fine-grained control of the replayed events and flexible post-attack analysis. We characterized nondeterministic events (those that need to be logged to reproduce the run) in terms of architecture level events: hardware interrupts and input events from external devices. In our evaluations, our approach imposed low performance overhead in terms of performance and log file growth rate.

Following this, we applied this replay approach on post-attack recovery from control-flow hijacking Internet worms. The main issue in this work was how we could maintain the availability of a host after an attack where much of the system state could have gone corrupted. Basically we tracked down the source of network bytes in the system and, after an attack, we replayed the checkpointed run while ignoring malicious inputs. The challenge of this work was replaying the execution while removing malicious events. There had been some previous work in this area, but this work was the first to address the issue of shared, corrupted state. Our recovery strategy was successful when tested against several notorious Internet worms. The main runtime overhead of our approach was the cost of keeping track of low integrity bytes at the architecture level (VM).

In the process of designing and implementing our post-attack recovery approach we faced some challenges and limitations that were due to the semantic gap between the operating system layer and the architecture layer, for example, the change in the system entropy during recovery affected the behavior of the TCP server causing some connections to be reset. A security solution operating only at the VM level and aiming to preserve the integrity of an OS code and data against attacks can access only architecture-level objects, such as CPU, physical memory, registers and external devices. However, in order to protect the OS kernel, such solution will need to manipulate OS-level objects such as processes, files and allocated memory areas.

To bridge this semantic gap we noticed that we needed strong collaboration between the OS and the architecture layer (represented by a VM) and we also needed that the OS and VM layers were aware of the existence of one another. This was very different from the traditional VM usage model adopted, in its various forms, by the majority of current security solutions. In this traditional model, the OS is unaware of the presence of a virtualization layer, which does not count on the guest OS to enhance system security. To step towards bridging this semantic gap, we proposed an architecture of collaboration between a VM layer and the guest OS. In this architecture the OS could down-call the VM by issuing specific requests through an unused software interrupt in the Intel x86 architecture. The VM layer interacted with the OS layer through exceptions. To validate our model we applied it on a case study to protect the OS kernel code and data against malicious tampering. Protecting the kernel is a difficult problem given its complexity and variety and number of its data structures. We have chosen this case study because the majority of solutions to protect the OS kernel involved only detection and most of them were not effective against non-control data attacks, *i.e.*, attacks that do not rely on changing the kernel code to be effective. Also, research works proposing prevention mechanisms usually relied on code authentication or security policies that may not address all possible data structures that can be abused by an attacker. In this specific case study we employed a relaxed version of Biba's integrity model where no suspicious or low integrity subjects (active entities, such as kernel threads and functions and instructions at the VM level) are allowed to write into trusted or high integrity objects (passive entities, such as files and memory areas). To preserve kernel integrity, we kept track of low integrity data at the

OS and architecture level. The OS down-called the VM at certain points during system execution to request information about the integrity level of VM objects, and to pass information to the VM regarding memory locations, boundaries of data structures and implicit propagation of low integrity data. We enforced this integrity model by having the VM check the validity of all attempted writes into kernel code and data. In our experiments (with six kernel rootkits) all violations were prevented and the main runtime overhead of our approach was the cost of keeping track of low integrity bytes at the architecture level.

6.2 Extensions and Enhancements

Each of the chapters of this dissertation addresses specifics problems and, in each one, we propose mechanisms to address the problem and validate these mechanisms through proof-of-concept prototypes. The problems we have worked on raised many issues that I plan to address in the future. Also, for all the proposed approaches there are many possible extensions that can enhance them.

Our VM-based full system replay employed in attack analysis and recover can be extended for multiprocessors and DMA. It is currently targeted for uniprocessors because the version of the Intel x86 emulator [27] adopted as our VM layer did not support DMA. Another extension is to have checkpoints saved in a non-volatile medium. In our proof-of-concept prototype the checkpoint was represented by a duplicated Bochs process and filesystem redolog files.

In our recovery approach whenever we discovered an attack, we went back to the most recent checkpoint to recover the run. For all exploits tested during our evaluation this was sufficient, as the interval between attack injection and detection was very small (in the order of millions of instructions) compared to the checkpoint window. However, if the malicious input enters the system before the last checkpoint, our strategy will not be able to recover the host execution and clean it from the attack effects. One enhancement is to maintain several past checkpoints and before taking recovery actions, discover to which checkpoint to return. Related to this issue, another future work is to research for optimum checkpoint intervals. Although the exploits used to validate our approach involved one particular TCP connection, we are also interested in correlating different attack sources when an exploit is multi-staged and uses different TCP connections.

The recovery approach proposed in this research showed that we need a mechanism to control and reproduce system entropy during the recovery through replay phase. During recovery, we removed all the malicious events from the execution trace, thus changing system entropy. As a result, all the TCP connections started during the small recovery window were lost due to mismatched acknowledgement and sequence numbers (because the system entropy was changed) seen by the TCP server. Just logging the TCP sequence numbers did not solve the problem as the timing and even the size of the packets sent by the TCP server changed when system entropy changed. The challenge is how to keep or reproduce the entropy generated by malicious activities during recovery, while still preventing the malicious events from taking place. We plan to use collaboration between the guest OS and the VM to maintain the system entropy the same, while still ignoring the effects of malicious nondeterministic events in the system.

Further, the memory monitor component we have introduced in the virtual machine layer to perform data flow tracking incurs significant performance penalty because, for every instruction executed, the tags associated with the operands need to be checked and possibly propagated to a certain memory address or register. I plan to investigate how a multicore architecture could be used to improve data flow tracking performance.

For the collaborative architecture between OS-VM we would like to demonstrate how this model can be successfully employed in other types of security solutions, especially those involving other aspects of computer security such as confidentiality and availability. In the case study described in this dissertation our challenge was to preserve the OS code and data integrity.

6.3 Outlook and Contributions

This dissertation has made two important contributions to the field. First, we showed the importance of shifting the focus from only detection approaches, to prevention and response in the area of host security. And we demonstrated this with mechanisms and three proof-of-concept prototypes. As discussed in Chapter 1, most of the works in the literature regarding security focus only on detection approaches and detection alone is inefficient for countering the type of malware and attacks we are going to face in the next decade. We are facing a new generation of malware and attacker's motivations. New threats are spreading slower than their predecessors to avoid detection and malware, when found in a host, is usually not alone; the host is probably already infected with several unrelated malicious software and rootkits. Availability is increasingly important because many servers do not tolerate performance degradation or long downtimes, and in many countries, critical utilities such as water and the electrical grid are controlled via networked computer systems. In the next decade, businesses that can continue functioning in spite of an attack will have an edge compared to those that remain unavailable.

Second, we proposed collaboration between the OS and the architecture layer (represented by a VM) as an effective approach to build stronger security solutions and introduced a new VM usage model based on collaboration between a virtualization-aware OS and a VM layer. Collaboration helps us bridge the semantic gap between software and architecture layers, which is the root of many shortcomings in current security solutions. We showed, in the case study to protect OS kernel code and data against malicious tampering, that collaboration allows for better system protection. There is room for further research on the potential of collaboration between layers of software and architecture. Although in my current research I began addressing this issue of collaboration between these layers, I believe that I barely scratched the surface of the problem and research possibilities. Several issues remain to be addressed and solved, such as, how we can extend this collaboration model to other layers and how we can address aspects of confidentiality and availability through collaboration.

The insights gained during the research described in this dissertation will interest other researchers and developers seeking to devise strong, flexible and fine-grained security approaches. The work is promising because this model of collaboration between layers can be applied not only in several aspects of computer security, but also in other settings as well.

BIBLIOGRAPHY

[1] Microsoft. Microsoft Security Intelligence Report - January through June 2009, 2009.

[2] Sans. Sans Top-20 2009 Security Risks - Annual Update, 2009.

[3] McAfee. McAfee Virtual Criminology Report - 2009.

[4] NITRD Senior Steering Group on Cybersecurity R&D. National Cyber R&D Framework: Changing The Game. *Annual Computer Security Applications Conference (ACSAC)*, December 2009.

[5] Matt Bishop. *Computer Security: Art and Science.* 2003. ISBN 0-201-44099-7.

[6] Peter Szor. *The Art of Computer Virus Research and Defense.* 2005. ISBN 0-321-30454-3.

[7] Jedidiah Richard Crandall. *Capturing and Analyzing Internet Worms.* PhD thesis, Department of Computer Science - University of California at Davis, June 2007.

[8] Daniela Oliveira, Jedidiah Crandall, Gary Wassermann, S. Felix Wu, Zhendong Su, and Frederic Chong. ExecRecorder: VM-Based Full-System Replay for Attack Analysis and System Recovery. *Workshop on Architectural and System Support for Improving Software Dependability - ASID'06 (with ASPLOS 2006)*, pages 66–71, October 2006.

[9] E. N. Elnozahy, Lorenzo Alvisi, Yi-Min Wang, and David B. Johnson. A Survey of Rollback-Recovery Protocols in Message-Passing Systems. *University of Michigan Technical Report CSE-TR-410*, 31(3):375–408, September 2002.

[10] L. Alvisi. *Understanding the Message Logging Paradigm for Masking Process Crashes.* PhD thesis, Cornell University, 1996.

[11] J. H. Slye and E. N. Elnozahy. Support for Software Interrupts in Log-Based Rollback-Recovery. *IEEE Transactions on Computers*, 47(10):1113–1123, October 1998.

[12] David E. Lowell and Peter M. Chen. Discount Checking: Transparent, Low-Overhead Recovery for General Applications. *University of Michigan Technical Report CSE-TR-410-99*, 1998.

[13] Thomas C. Bressoud and Fred B. Schneider. Hypervisor-Based Fault Tolerance. *ACM TOCS*, 14(1):80–107, February 1996.

[14] L. Alvisi and K. Marzullo. Message Logging: Pessimistic, Optimistic, Causal, and Optimal. *IEEE Transactions on Software Engineering*, 24(2):149–159, February 1998.

[15] J.H. Slye and E.N. Elnozahy. Supporting Nondeterministic Execution in Fault-Tolerant Systems. *FTCS*, 1996.

[16] George W. Dunlap, Samuel T. King, Sukru Cinar, Murtaza A. Basrai, and Peter M. Chen. ReVirt: Enabling Intrusion Analysis through Virtual-Machine Logging and Replay. *SIGOPS Oper. Syst. Rev.*, 36(SI):211–224, 2002. ISSN 0163-5980. doi: http://doi.acm.org/10.1145/844128.844148.

[17] Zvi Gutterman and Benny Pinkas. BugNet: Continuously Recording Program Execution for Deterministic Replay Debugging. *ISCA-32*, pages 284–295, June 2005.

[18] F. Qin, J. Tucek, J. Sundaresan, and Y. Zhou. Rx: Treating Bugs as Allergies - A Safe Method to Survive Software Failures. *ACM SOSP*, pages 235–248, October 2005.

[19] Sudarshan M. Srinivasan, Srikanth Kandula, Christopher R. Andrews, and Yuanyuan Zhou. Flashback: A Lightweight Extension for Rollback and Deterministic Replay for Software Debugging. *USENIX*, June 2004.

[20] T. J. LeBlanc and J. M. Mellor-Crummey. Debugging Parallel Programs with Instant Replay. *IEEE Transactions on Computers*, 36(4):471–482, April 1987.

[21] Jong-Deok Choi and Harini Srinivasan. Deterministic Replay of Java Multithreaded Applications. *ACM SIGMETRICS SPDT*, pages 48–59, August 1998.

[22] M. Prvulovic and J. Torrellas. ReEnact: Using Thread-Level Speculation Mechanisms to Debug Data Races in Multithreaded Codes. *ISCA-30*, pages 110–121, June 2003.

[23] M. Xu, R. Bodik, and M. D. Hil. A Flight Data Recorder for Enabling Full-System Multiprocessor Deterministic Replay. *ISCA-30*, pages 122–133, June 2003.

[24] Jedidiah R. Crandall and Frederic T. Chong. Minos: Control Data Attack Prevention Orthogonal to Memory Model. *MICRO*, pages 221–232, December 2004.

[25] Jedidiah R. Crandall, Zhendong Su, S. Felix Wu, and Frederic T. Chong. On Deriving Unknown Vulnerabilities from Zero-Day Polymorphic and Metamorphic Worm Exploits. *ACM CCS*, pages 235–248, November 2005. URL http://wwwcsif.cs.ucdavis.edu/~crandall/ccsdacoda.pdf.

[26] Robert Love. *Linux Kernel Development*. Novell Press, 2005.

[27] Bochs. bochs: the Open Source IA-32 Emulation Project (http://bochs.sourceforge.net).

[28] Intel Corporation. IA-32 Intel Architecture Software Developer's Manual. Volume 1: Basic Architecture. 2001.

[29] Web benchmark. http://www.serverwatch.com/news/article.php/10824_1133391_2.

[30] UnixBench (http://www.tux.org/pub/tux/benchmarks/).

[31] Microsoft. Microsoft SQLIO. http://www.microsoft.com/downloads/.

[32] Ashlesha Joshi, Samuel T. King, George W. Dunlap, and Peter M. Chen. Detecting Past and Present Intrusions through Vulnerability-specific Predicates. *ACM SOSP*, pages 91–101, October 2005. URL http://www.cs.washington.edu/homes/gribble/papers/IEEE_vmm.pdf.

[33] Inetcop wu-ftpd 2.6.0 Vulnerability (http://x82.inetcop.org/h0ne/papers/free-ur-mind.pdf).

[34] Jedidiah R. Crandall, John Brevik, Shaozhi Ye, Gary Wassermann, Daniela A. S. de Oliveira, Zhendong Su, S. Felix Wu, and Frederic T. Chong. Putting Trojans on the Horns of a Dilemma: Redundancy for Information Theft Detection. *Special Issue on Security in Computing of the Transactions on Computational Sciences Journal (Springer LNCS)*, 2008.

[35] Daniela Oliveira, Jedidiah Crandall, Gary Wassermann, Shaozhi Ye, S. Felix Wu, Zhendong Su, and Frederic Chong. Bezoar: Automated Virtual Machine-based Full-System Recovery from Control-Flow Hijacking Attacks. *IEEE/IFIP Network Operations and Management Symposium (NOMS)*, April 2008.

[36] P. A. Lee and T. Anderson. *Fault Tolerance Principles and Practice 2nd. ed.* Springer-Verlag Wien New York, 1990.

[37] James Newsome and Dawn Song. Dynamic taint analysis for automatic detection, analysis, and signature generation of exploits on commodity software. In *NDSS*, February 2005.

[38] Feng Qin, Cheng Wang, Zhenmin Li, Ho-seop Kim, Yuanyuan Zhou, and Youfeng Wu. LIFT: A Low-Overhead Practical Information Flow Tracking System for Detecting Security Attacks. *MICRO-39*, pages 135–148, December 2006.

[39] Georgios Portokalidis, Asia Slowinska, and Herbert Bos. Argos: an Emulator for Fingerprinting Zero-Day Attacks. *EuroSys*, April 2006.

[10] S. Bhatkar, D. DuVarney, and R. Sekar. Address obfuscation: An efficient approach to combat a broad range of memory error exploits. *USENIX Security*, 2003.

[11] PAX. The PAX team. http://pax.grsecurity.net.

[12] George C. Necula, Scott McPeak, and Westley Weimer. CCured: type-safe retrofitting of legacy code. In *POPL*, pages 128–139, 2002. URL citeseer.nj.nec.com/necula02ccured.html.

[13] T. Jim, Greg Morrisett, Dan Grossman, Michael Hicks, James Cheney, and Yanling Wang. Cyclone: a safe dialect of C. *USENIX*, 2002.

[44] C. Cowan, C. Pu, D. Maier, J. Walpole, P. Bakke, S. Beattie, A. Grier, P. Wagle, Q. Zhang, and H. Hinton. StackGuard: Automatic adaptive detection and prevention of buffer-overflow attacks. In *USENIX Security*, pages 63–78, Jan 1998. URL http://www.usenix.org/publications/library/proceedings/sec98/full_papers/cowan/cowan.pdf.

[45] Jun Xu, Peng Ning, Chongkyung Kil, Yan Zhai, and Chris Bookholt. Automatic Diagnosis and Response to Memory Corruption Vulnerabilities. *ACM CCS*, November 2005.

[16] Zhenkai Liang and R. Sekar. Fast and Automated Generation of Attack Signatures: A Basis for Building Self-Protected Servers. *ACM CCS*, November 2005.

[17] David Brumley, James Newsome, Dawn Song, Hao Wang, and Somesh Jha. Towards Automatic Generation of Vulnerability-Based Signatures. *IEEE Symposium on Security and Privacy*, May 2006.

[18] Manuel Costa, Jon Crowcroft, Miguel Castro, Antony Rowstron, Lidong Zhou, Lintao Zhang, and Paul Barham. Vigilante: End-to-end containment of Internet worms. In *ACM SOSP*, 2005.

[19] Ryan Iwahashi, Daniela Oliveira, S. Felix Wu, Jedidiah Crandall, Young-Jun Heo, Jin-Tae Oh, and Jong-Soo Jang. Towards Automatically Generating

Double-Free Vulnerability Signatures Using Petri Nets. *Information Security Conference (ISC)*, September 2008.

[50] Joseph Tucek, James Newsome, Shan Lu, Chengdu Huang, Spiros Xanthos, David Brumley, Yuanyuan Zhou, and Dawn Song. Sweeper: A Lightweight End-to-End System for Defending Against Fast Worms. *EuroSys*, March 2007.

[51] Weidong Shi, Hsien-Hsin S. Lee, Laura Falk, and Mrinmoy Ghosh. An Integrated Framework for Dependable and Revivable Architectures Using Multicore Processor. *ISCA-33*, 34(2):102–113, May 2006.

[52] Stelios Sidiroglou, Oren Laadan, Angelos D. Keromytis, and Jason Nieh. Using Rescue Points to Navigate Software Recovery. *IEEE Symposium on Security & Privacy*, May 2007.

[53] Stelios Sidiroglou, Michael E. Locasto, Stephen W. Boyd, and Angelos D. Keromytis. Building a Reactive Immune System for Software Services. *USENIX*, April 2005.

[54] Alexey Smirnov and Tzi-cker Chiueh. DIRA: Automatic Detection, Identification, and Repair of Control-Hijacking Attacks. *NDSS*, February 2005.

[55] Martin Rinard, Cristian Cadar, Daniel Dumitran, Daniel M. Roy, Tudor Leu, and Jr. William S. Beebee. Enhancing Server Availability and Security through Failure-Oblivious Computing. *OSDI*, 2004.

[56] B. Jack. Remote Windows Kernel Exploitation - Step into the Ring 0. *eEye Digital Security Whitepaper*, 2005.

[57] M. Kharbutli, X. Jiang, Y. Solihin, G. Venkataramani, and M. Prvulovic. Comprehensively and Efficiently Protecting the Heap. *ASPLOS*, October 2006.

[58] Security Focus. Security Focus Vulnerability Notes, bugtraq id 1316. URL http://www.securityfocus.com/bid/1316/discussion/.

[59] K. J. Biba. Integrity Considerations for Secure Computer Systems. In *MITRE Technical Report TR-3153*, Apr 1977.

[60] Alex Ho, Michael Fetterman, Christopher Clark, Andrew Warfield, and Steven Hand. Practical taint-based protection using demand emulation. *EuroSys*, 2006.

[61] Aaron B. Brown and David A. Patterson. Rewind, Repair, Replay: Three R's to Dependability. *10th ACM SIGOPS European Workshop*, pages 70–77, 2002.

[62] Security Focus. Security Focus Vulnerability Notes, (http://www.securityfocus.com). bid == Bugtraq ID. URL http://www.securityfocus.com/bid/3707/discussion/.

[63] Webstone. Webstone 2.5. http://www.mindcraft.com/webstone/.

[64] Daniela Oliveira and S. Felix Wu. Protecting Kernel Code and Data with a Virtualization-Aware Collaborative Operating System. *Annual Computer Security Applications Conference (ACSAC)*, December 2009.

[65] James E. Smith and Ravi Nair. *Virtual Machines - Versatile Platforms for Systems and Processes*. Morgan Kaufmann, 2005.

[66] Peter M. Chen and Brian D. Noble. When Virtual is Better than Real. *HotOS*, May 2001. URL http://www.eecs.umich.edu/Rio/papers/chen01.pdf.

[67] Tal Garfinkel and Mendel Rosenblum. A Virtual Machine Introspection Based Architecture for Intrusion Detection. *Network and Distributed System Security Symposium*, 2003. URL http://suif.stanford.edu/papers/vmi-ndss03.pdf.

[68] Yi-Min Wang, Doug Beck, Xuxian Jiang, Roussi Roussev, Chad Verbowski, Shuo Chen, and Sam King. Automated Web Patrol with Strider HoneyMonkeys: Finding Web Sites That Exploit Browser Vulnerabilities. *NDSS*, 2006.

[69] Arvind Seshadri, Mark Luk, Ning Qu, and Adrian Perrig. SecVisor: A Tiny Hypervisor to Provide Lifetime Kernel Code Integrity for Commodity OSes. *ACM Symposium on Operating Systems Principles (SOSP)*, October 2007.

[70] Stephen T. Jones, Andrea C. Arpaci-Dusseau, and Remzi H. Arpaci-Dusseau. VMM-based Hidden Process Detection and Identification using Lycosid. *ACM SIGPLAN/SIGOPS International Conference on Virtual Execution Environments*, 2008.

[71] Ryan Riley, Xuxian Jiang, and Dongyan Xu. Guest-Transparent Prevention of Kernel Rootkits with VMM-based Memory Shadowing. *RAID*, 2008.

[72] Tal Garfinkel, Ben Pfaff, Jim Chow, Mendel Rosenblum, and Dan Boneh. Terra: A Virtual Machine-Based Platform for Trusted Computing. *ACM Symposium on Operating Systems Principles*, pages 193–206, October 2003. URL http://www.cs.rochester.edu/sosp2003/papers/p199-garfinkel.pdf.

[73] Arati Baliga and Liviu Iftode. Automated Containment of Rootkit Attacks. *Computer and Security, Elsevier*, 2008.

[74] Min Xu, Xuxian Jiang, Ravi Sandhu, and Xinwen Zhang. Towards a VMM-based Usage Control Framework for OS Kernel Integrity Protection. *SACMAT*, 2007.

[75] Nick L. Petroni Jr. and Michael Hicks. Automated Detection of Persistent Kernel Control-Flow Attacks. *ACM CCS*, pages 103–115, November 2007.

[76] Lionel Litty, H. Lagar-Cavilla, and David Lie. Hypervisor support for identifying covertly executing binaries. *USENIX*, 2008.

[77] Bryan Payne, Martim Carbone, Monirul Sharif, and Wenke Lee. Lares: An Architecture for Secure Active Monitoring using Virtualization. *IEEE Symposium on Security and Privacy*, May 2008.

[78] Heng Yin, Dawn Song, Manuel Egele, Christopher Kruegel, and Engin Kirda. Panorama: Capturing System-wide Information Flow for Malware Detection and Analysis. *ACM CCS 07*, pages 116–127, November 2007.

[79] Asia Slowinska and Herbert Bos. The Age of Data: pinpointing guilty bytes in polymorphic buffer overflows on heap and stack. *ACSAC*, December 2007.

[80] Tal Garfinkel, Keith Adams, Andrew Warfield, and Jason Franklin. Compatibility is Not Transparency: VMM Detection Myths and Realities. *HotOS*, 2007.

[81] Samuel T. King, Peter M. Chen, Yi-Min Wang, Chad Verbowski, Helen J. Wang, and Jacob R. Lorch. SubVirt: Implementing malware with virtual machines. *IEEE Security and Privacy*, May 2006.

[82] Joanna Rutkowska. Subverting VistaTM Kernel For Fun And Profit. *Black Hat Briefings*, 2006.

[83] Peter Ferrie. Attacks on Virtual Machine Emulators. *Symantec Advanced Threat Research*, 2007.

[84] John Scott Robin and Cynthia E. Irvine. Analysis of the Intel Pentium's ability to support a secure virtual machine monitor. *USENIX*, 2000.

[85] Arati Baliga, Pandurang Kamat, and Liviu Iftode. Lurking in the Shadows: Identifying Systemic Threats to Kernel Data. *IEEE S&P'07*, pages 246–251, May 2007.

[86] LWN. Who needs /dev/kmem? (http://lwn.net/Articles/147901/). URL http://lwn.net/Articles/147901/.

[87] Francis M. David, Ellick M. Chan, Jeffrey C. Carlyle, and Roy H. Campbell. Cloaker: Hardware Supported Rootkit Concealment. *IEEE Security and Privacy*, pages 296–310, 2008.

[88] Xuxian Jiang, Xinyuan Wang, and Dongyan Xu. Stealthy malware detection through vmm-based "out-of-the-box" semantic view reconstruction. *ACM CCS*, pages 128–138, November 2007.

[89] Nick L. Petroni Jr., Timothy Fraser, and William A. Arbaugh. Copilot - a Coprocessor-based Kernel Runtime Integrity Monitor. *USENIX*, 2004.

[90] Nick L. Petroni Jr., Timothy Fraser, Aaron Walters, and William A. Arbaugh. An Architecture for Specification-Based Detection of Semantic Integrity Violations in Kernel Dynamic Data. *USENIX Security*, 2006.

[91] Yi-Min Wang, Doug Beck, Binh Vo, Roussi Roussev, and Chad Verbowski. Detecting Stealth Software with Strider GhostBuster. *DSN*, 2005.

[92] Arati Baliga, Vinod Ganapathy, and Liviu Iftode. Automatic Inference and Enforcement of Kernel Data Structure Invariants. *Annual Computer Security Applications Conference (ACSAC)*, pages 77–86, December 2008.

[93] Intel Corporation. IA-32 Intel Architecture Software Developer's Manual. Volume 2: Instruction Set Reference. 2001.

[94] Daniel P. Bovet and Marco Cesati. *Understanding the Linux kernel*. O'Reilly, 2005.

[95] IBM. GCC Hacks in the Linux Kernel (http://www.ibm.com/developerworks/linux/library/l-gcc-hacks/). URL http://www.ibm.com/developerworks/linux/library/l-gcc-hacks/.

[96] S. Chen, J. Xu, and E. C. Sezer. Non-control-hijacking attacks are realistic threats. In *USENIX Security Symposium*, 2005.

[97] Bughunter. Linux on-the-fly kernel patching without LKM (http://doc.bughunter.net/rootkit-backdoor/kernel-patching.html).

[98] SourceForge.net: Open Source Software (http://sourceforge.net).

[99] SPEC - Standard Performance Evaluation Corporation (http://ftp.spec.org/cpu2000/CINT2000).

[100] Jinpeng Wei, Bryan D. Payne, Jonathon Giffin, and Calton Pu. Soft-Timer Driven Transient Kernel Control Flow Attacks and Defense. *Annual Computer Security Applications Conference (ACSAC)*, 2008.

[101] Samuel T. King, George W. Dunlap, and Peter M. Chen. Debugging Operating Systems with Time-Traveling Virtual Machines. *Proceedings of the 2005 USENIX Technical Conference*, April 2005.

[102] Tal Garfinkel and Mendel Rosenblum. When Virtual is Harder than Real: Security Challenges in Virtual Machine Based Computing Environments. *HotOS*, June 2005. URL http://www.usenix.org/events/hotos05/prelim_papers/garfinkel/garfinkel.pdf.

[103] Timothy Roscoe, Kevin Elphinstone, and Gernot Heiser. Hype and Virtue. *HotOS*, 2007.

[104] Intel Corporation. Vanderpool Technology - Technical Report. 2005.

[105] AMD. AMD Pacifica Virtualization Technology.

[106] Lionel Litty and David Lie. Manitou: a layer-below approach to fighting malware. *ASID*, October 2006.

[107] Julian Bennett Grizzard. *Towards Self-Healing Systems: Re-establishing Trust in Compromised Systems*. PhD thesis, Georgia Institute of Technology, May 2006.

[108] Kevin R. B. Butler, Stephen McLaughlin, and Patrick D. McDaniel. Rootkit-Resistant Disks. *ACM CCS*, pages 403–416, November 2008.

[109] John Criswell, Andrew Lenharth, Dinakar Dhurjati, and Vikram Adve. Secure Virtual Architecture: A Safe Execution Environment for Commodity Operating Systems. *SOSP*, October 2007.

[110] Rick Kennell and Leah H. Jamieson. Establishing the genuinity of remote computer systems. *USENIX*, August 2003.

[111] Reiner Sailer, Trent Jaeger, Xiaolan Zhang, and Leendert van Doorn. Attestation-based policy enforcement for remote access. *ACM CCS*, November 2004.

[112] Reiner Sailer, Trent Jaeger, Xiaolan Zhang, and Leendert van Doorn. Design and implementation of a TCG-based integrity measurement architecture. *USENIX*, August 2004.

[113] M. Abadi, M. Budiu, U. Erlingsson, and J. Ligatti. Control-flow Integrity. *ACM CCS*, November 2005.

[114] Zhi Wang, Xuxian Jiang, Weidong Cui, and Xinyuan Wang. Countering Persistent Kernel Rootkits Through Systematic Hook Discovery. *RAID*, September 2008.

[115] G. Edward Suh, Jaewook Lee, and Srinivas Devadas. Secure Program Execution via Dynamic Information Flow Tracking. In *Proceedings of ASPLOS-XI*, October 2004.

[116] Jim Chow, Ben Pfaff, Tal Garfinkel, and Mendel Rosenblum. Understanding Data Lifetime via Whole System Simulation. *USENIX*, 2004.

[117] Philipp Vogt, Florian Nentwich, Nenad Jovanovic, Engin Kirda, Christopher Kruegel, and Giovanni Vigna. Cross-Site Scripting Prevention with Dynamic Data Tainting and Static Analysis. *NDSS*, 2007.

[118] Haibo Chen, Xi Wu, Liwei Yuan, Binyu Zang, Pen chung Yew, and Frederic T. Chong. From Speculation to Security: Practical and Efficient Information Flow Tracking Using Speculative Hardware. *ISCA*, June 2008.

[119] Walter Chang, Brandon Streiff, and Calvin Lin. Efficient and Extensible Security Enforcement Using Dynamic Data Flow Analysis. *ACM CCS*, November 2008.

[120] Phillip A. Bernstein, Vassos Hadzilacos, and Nathan Goodman. *Concurrency Control and Recovery in Database Systems*. Springer-Verlag Wien New York, 1987.

[121] T. C. Chou. Beyond fault tolerance. *IEEE Computer*, 30(4):31–36, 1997.

[122] J. Gray. Why do computers stop and what can be done about it? *5th Symp. on Reliability in Distributed Software and Database Systems*, 1986.

[123] M. Sullivan and R. Chillarege. Software Defects and their Impact on System Availability - A Study of Field Failures in Operating Systems. *International Symposium on Fault-Tolerant Computing*, 1991.

[124] Mary Baker and Mark Sullivan. The Recovery Box: Using Fast Recovery to Provide High Availability in the UNIX Environment. *USENIX*, June 1992.

[125] George Candea, Shinichi Kawamoto, Yuichi Fujiki, Greg Friedman, and Armando Fox. Microreboot – A Technique for Cheap Recovery. *OSDI*, December 2004.

[126] Y. Huang, C. Kintala, N. Kolettis, and N. Fulton. Software Rejuvenation: Analysis, Module and Applications. *FTCS-25*, pages 381–390, June 1995.

[127] James Newsome, David Brumley, Jason Franklin, and Dawn Song. Replayer: Automatic Protocol Replay by Binary Analysis. *ACM CCS*, November 2006.

[128] Weidong Cui, Vern Paxson, Nicholas Weaver, and Randy H. Katz. Protocol-Independent Adaptive Replay of Application Dialog. *NDSS*, February 2006.

[129] F. Sultan, A. Bohra, P. Gallard, I. Neamtiu, S. Smaldone, Y. Pan, and L. Iftode. Recovering Internet Service Sessions from Operating System Failures. *IEEE Internet Computing*, 9(2):17–27, March/April 2005.

[130] Emery D. Berger and Benjamin G. Zorn. DieHard: Probabilistic Memory Safety for Unsafe Languages. *PLDI*, pages 158–168, June 2006.

[131] Gene Novark, Emery D. Berger, and Benjamin G. Zorn. Exterminator: Automatically Correcting Memory Errors with High Probability. *PLDI*, pages 158–168, June 2007.

[132] Ashvin Goel, Kenneth Po, Kamran Farhadi, Zheng Li, and Eyal de Lara. The taser intrusion recovery system. *ACM SOSP*, pages 163–176, 2005.

[133] Francis Hsu, Hao Chen, Thomas Ristenpart, Jason Li, and Zhendong Su. Back to the Future: A Framework for Automatic Malware Removal and System Repair. *ACSAC*, pages 163-176, December 2006.

CPSIA information can be obtained at www.ICGtesting.com
Printed in the USA
BVOW10s1446150114

342004BV00010B/368/P